THE NEW COMMUNES

RON E. ROBERTS

is Assistant Professor of Sociology at the University of Northern Iowa. He has traveled extensively in the United States, interviewing communalists of all beliefs and persuasions.

THE NEW COMMUNES

Coming Together in America

RON E. ROBERTS

Prentice-Hall, Inc.

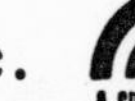

Englewood Cliffs, N. J.

A SPECTRUM BOOK

10 9 8 7 6 5 4

P: 0 13-612465-8; C: 0 13-612473-9
Library of Congress Catalog Card Number: 70-160531

Printed in the United States of America

Prentice-Hall International, Inc. (*London*)
Prentice-Hall of Australia, Pty. Ltd. (*Sydney*)
Prentice-Hall of Canada, Ltd. (*Toronto*)
Prentice-Hall of India Private Limited (*New Delhi*)
Prentice-Hall of Japan, Inc. (*Tokyo*)

To Professors Vernon Parenton and Rudolf Heberle
of Louisiana State University,
sociologists and humanists

PREFACE

"Let us dream daily, and with/extreme energy, but always keeping in mind that our/dreams will come to nothing. Let us be ardent and skeptical."

MAURICE BARRÈS

My intrigue with the new utopians began in two ways: I developed a generalized interest in utopian thought as I studied young people who were enmeshed in the utopian ideology of a rather orthodox religious sect—the Latter-Day Saints[1]—and my political concerns led me to follow with interest the fortunes of the New Left on college campuses. Whether one followed the moderate route of Gene McCarthy's "children's crusade" or the rather more militant voice of the Student Nonviolent Coordinating Committee or Students for a Democratic Society, the political climate of America brought tremendous frustration and anger to those seeking racial justice or an end to the seemingly interminable war in Vietnam. The apolitical "hippies," "diggers," and "heads" who attempted early semicommunal ventures in Greenwich Village or Haight-Ashbury had always maintained that the American society was not worth saving. Many later communalists, however, "dropped in" to their new utopias to gain strength to confront, rather than ignore, the outside world.

I have attempted in this book to describe some of the utopian and communalist movements now extant in the United States. Describing many of the groups is much like discussing the structure of the proverbial can of worms. Some utopian groups never get to the point of actualizing their ideas; others begin energetically only to fold up their tents (sometimes quite literally) and move on to new experiments. Also, since my definitions of utopians and communalists are inclusive, groups as dissimilar as California hip communalists and the Catholic workers are

1. See my unpublished doctoral dissertation, *Dilemmas of Utopian Commitment in a Contemporary Religious Sect* (Louisiana State University, 1969)

discussed. It is my hope that the underlying themes common to all these groups as well as their distinguishing traits will become evident to some degree in this book.

Many people have aided me in the preparation of this book, especially my intrepid fellow commune hoppers, Nic and Alan, Kirk and Eric, Patti and Chere and Johnsk.

Also I am grateful for the funds granted me by the research committee of the University of Northern Iowa for travel and clerical work. Linda Medland did a stalwart and excellent job of typing and editing the manuscript.

Thanks are also due Dick Fairfield of the *Modern Utopian*, Griscom and Jane Morgan of the Vale, Charles Davis of the Bhoodan Center, Patsy (Richardson) Sun of Freefolk, Kathy A. who shared her diary of her life at "Oz," and the staff of *Alternate Society* (a Canadian communalist magazine).

I would like also to thank the hundreds of communalists I interviewed in person or by letter. I am thankful, of course, for their information but more than this I am grateful for the opportunity I had to share with them. We did share. I, on my part, donated two comic books, gasoline, and cigarette money to hip communalists at Wheelers Free Ranch. From communalists at Morningstar, my friends and I received apples, pears, and an ornately carved walking stick. Later, at Wheeler's Ranch "freeks" and friends exerted great efforts to get my disabled car moving again. I thank them all.

As to my own views toward the social experimentation described in this book, they are in the main favorable. It is a commonplace that many of the young have lost faith in the institutions built by their fathers. Daniel Boorstin, a liberal critic of the New Left, has criticized today's radicals for their lack of programs and communities.[2] Had Boorstin searched past the hysteria of S.D.S., Weathermen, *et al.*, he might have found a number of radical young people creating programs and attempting to create communities.

Even if (as one would suppose) most current utopists fail in their immediate communal ventures, society will be no worse off than before. If they succeed even partially in their objectives, they could become the vanguard of a new age. It is, therefore, the duty of the concerned social scientist to consider the intellectual premises as well as the internal dynamics of modern utopian movements.

2. Daniel Boorstin, *The Decline of Radicalism: Reflections on America Today* (New York: Random House, 1970)

THE NEW COMMUNES

CONTENTS

1

IMAGES OF DISSENT, OPTIONS FOR CHANGE

"We have met the enemy and they are us."

Pogo

THE NEW COMMUNALISTS

Young people in America are attempting to create an alternative society. One of the stock arguments in the intergenerational warfare between the rebellious young[1] and the frightened and wrathful elder generation is that the young are critical of establishment institutions "but have nothing better to replace them with." It is true that a minute fraction of the young have been so pathologically angered by the melange of hypocracy, militarism, and manipulation in America that they have resorted to individual acts of terror, such as bombing or other forms of violence, with little regard for the future.

Of those who dissent from the views of the "silent majority," [2] however, an increasing number *are* able to respond with "something better." There are many paths to "something better" and it would be a disservice to truth to discuss only one of them. Of the many paths the young have chosen, all seem to adhere to Thoreau's advice ". . . simplify, simplify."

To simplify human relations, to make them more satisfying, great

1. I do not mean to infer at this point that most of the young are dissenters. Quite the contrary: a number of studies have shown that a higher proportion of the under thirty generation were more sympathetic to George Wallace than were their elders. See, for example, Seymour Lipset and Earl Raab, "The Wallace Whitelash," *Transaction* 17, no. 2 (December, 1969): 27.

Those most likely to dissent are of course the individuals who come from a background of affluence and who have come to see the internal contradictions in the larger society through their contact with higher education.

2. President Nixon's phrase for those who did not oppose his actions in Vietnam.

numbers of the young have rejected technological solutions to their problems. They have almost casually rejected the eternal optimism of liberalism along with the sterile conformism of current day conservatism. In a word, many of the young wish to be "retribalized." They wish to be rid of the neurotic attention given to economics and property in modern America. The Diggers, an early hip communalist group, established their unofficial headquarters in San Francisco and with a theatrical flourish began to establish "free stores." Begging for and giving away food, clothing, and money, the Diggers at once performed an act of solidarity with the hip community and a symbolic rejection of the money/property obsessed "straight" society. One of the leaders of the movement expressed both his diagnosis and prescription for the new order of things in this way:

> The Diggers are hip to property. Everything is free, do your own thing. Human beings are the means of exchange. Food, machines, clothing, materials, shelter and props are simply there. Stuff. A perfect dispenser would be an open automat on the street. Locks are time-consuming. Combinations are clocks.
>
> So a store of goods or clinic or restaurant that is free becomes a social art form. Ticketless theater. Out of money and control . . . Diggers assume free stores to liberate human nature. First free the space, goods, and services. Let theories of economics follow social facts. Once a free store is assumed, human wanting and giving, needing and taking become wide open to improvisation.[3]

The Diggers and the new communalists who followed them have had as their antecedents other social and intellectual movements which in many ways influenced and shaped their beliefs. Perhaps the two most influential precommunalist movements of recent times revolved around the "beat generation" of the late 1950s and the civil rights movement of the early 1960s.

COMMUNAL ANTECEDENTS—THE BEAT GENERATION

The beat generation was more in the nature of a literary rather than a social movement. It did provide, however, an image of dissent for young disaffected literati such as Ginsberg, Kerouac, Ferlinghetti, and Snyder. The symbolism used by beat writers and poets was pure "coun-

3. *The Realist*, no. 81 (August, 1968), p. 3.

terculture." Drugs, cool jazz, interracial sex, and Zen Buddhism were means of establishing a rejection of all things "one hundred percent American."

Pour épater le bourgeoisie[4] became a way of life for the beats. The conformity demanded by the larger American community was rejected by the beats advocacy of artistic integrity, voluntary poverty, and social disengagement. It was Allen Ginsberg who saw the best minds of his generation "destroyed by madness." More importantly, however, it was Jack Kerouac who described the social models who were to typify the beat rebellion. The model Kerouac chose for the central protagonist of his short stories and novels was the self-appointed outcast—the perpetual stranger. Kerouac's major works, *On the Road,*[5] *The Subterraneans,*[6] and his shorter "October in the Railroad Earth"[7] typify the sense of detachment and perpetual movement associated with the beat movement. Kerouac's characters hitchhiked around the country, made love in a casual way and generally refused to commit themselves to a sense of community with any group. It was, of course, this very romantic individualism which (together with the changing tides of fashion) provided the demise of the movement. Movements, trends, or fashions cannot be sustained without the aid of a community of support. Since the beat generation specifically rejected the idea of community, the life span of the movement was breathtakingly short.[8] This is not to say that the movement was without influence. Residues of the images of dissent proposed by the beats are still part of the hip culture.

COMMUNAL ANTECEDENTS—CIVIL RIGHTS ACTIVISTS

By the early 1960s significant numbers of upper middle class college students involved themselves in the struggle for civil rights in the deep south. It is impossible to overestimate the importance of the movement for those who were actively working in an organization such as the Student Nonviolent Coordinating Committee ("SNCC") in Mississippi in 1964. Essentially, two themes run through the biographies of

4. To shock or "put on" the bourgeoisie.
5. New York: Viking Press, 1957.
6. New York: Grove Press, 1958.
7. *Evergreen Review* 1, no. 2, pp. 119–36.
8. Save in the minds of some of the less discriminating public, who equated any form of nonconformity as beatnik. Hence early dissenters against the war in Vietnam were "peaceniks."

the black and white students who spent a summer or longer working on voter registration projects, the creation of "freedom schools," and the like.

The first theme is the realization that institutions promoting or maintaining the caste-like race relations in the South were not easily dislodged. Overt violence, the covert manipulations of Southern legislators, and simple fear, all contributed to the maintenance of what the students considered to be an essentially repressive society. The federal government, the students felt, was equivocal at best in initiating equalitarian changes in the interests of disinfranchised blacks. No less ambivalent was the rest of white America as a number of Northern newspapers questioned the purity of motives as well as the wisdom of the student ventures. In the main, the white and black civil rights workers left their involvement in the South with mixed feelings and only partial victories. It is true that a goodly amount of civil rights legislation was passed in part due to their efforts. Nevertheless, many young activists were of the opinion that the generally repressive nature of American society had been unmasked in the struggle.

Another latent effect of the early civil rights movement was the discovery by many white activists of "soul." Soul in the black community refers to a kind of authenticity—a "nonphony" experience. More than this, many whites experienced for the first time a real sense of community in the Southern Negro church. One upper middle class activist told me in 1964 that she had never felt "alive" until attending her first service in a rural, black Mississippi church.

Howard Zinn quotes a white girl from Virginia who expressed her experience with the movement in this way, ". . . finally it all boils down to human relationships. It has nothing to do finally with governments. It is the question of whether we—whether I shall go on living in isolation or whether there shall be a we. The student movement is not a cause . . . it is an I'm going to sit beside you." [9] This essentially nonpolitical statement is evidence of a real desire for community. Concern for the individual, mutuality of support, and a sense of belonging are part and parcel of the "psychological texture" [10] of communities.

In retrospect, then, one can see evidence of a real change in the images of dissent, from the beat generation of the 1950s to the committed

9. Quoted in Howard Zinn S.N.C.C., *The New Abolitionists* (Boston: Beacon Press, 1964), p. 7.
10. This term is taken from Rudolf Heberle.

community-oriented dissidents of the early civil rights movement.[11] A new kind of "reality principle" began to be implicitly accepted by those wishing to confront or even withdraw from bourgeois society. This was the fact that effective confrontation or even withdrawal from the larger society is best accomplished by community support. It is a sociological axiom that individuals as members of a community can endure great privation and difficulties.[12] Conversely, individuals without strong primary group ties are prone to a number of varieties of personality disorganization.

This is a lesson well learned by modern communalists, who in spite of their individualism and nonconformity have come to realize the necessity for building new and hopefully more humane forms of community.

FROM PRAXIS TO ANALYSIS

The question for the post civil rights dissidents of the late 1960s and early 1970s was not whether American society was repressive and dehumanizing (they had concluded that it was) but rather the question as to the nature of that repression and the best means for ending it. Before one can opt for a "free" society, he must develop a perspective concerning the nature of repression. Why are men "everywhere in chains"?

Materialism. One of the oldest answers to this question is that no society can be "free" without proper attention to spiritual matters. For this reason a number of religious communal organizations have chosen contemplation, ritual, or service to others as a means of self-awareness, hence self-liberation. Groups as diverse as the Ahimsa (nonviolence) community, a Buddhist group in Kansas, and Tivoli, a Catholic workers communal farm in New York, both represent this point of view.

Sexual Repression—The Monogamous Family. Some analysts of a neo-Freudian bent would regard it as no accident that social systems with a highly repressive sex code tend toward totalitarian politics. They

11. By "early civil rights movement" I refer to the period from about 1960, when sit-ins and "freedom rides" began to occur in the South, to 1966, when the "black power" concept (which considered integration irrelevant) came into being.

12. See for example Robert Coles, *Children of Crisis*, for an analysis of the emotional resources of a six-year-old who with the aid of her family endured a gauntlet of screaming whites daily to attend a newly integrated school in New Orleans. (New York: Dell Publishing Co., 1967.)

would cite the puritan's harsh early New England, Nazi Germany, and the current Republic of South Africa (which has officially banned the mini-skirt) as cases in point. If, indeed, authoritarianism is caused by the sexual possessiveness of the monogamous family, as many communalists believe, then radical changes in the male-female relationship may be requisite to healthy, fulfilling human existence. Therefore, a number of modern communalists regard the *lack* of a sexual-familial revolution in America as a factor in the increasing rise of violence and political repression in the nation. "Make love not war" may be a hackneyed phrase but for those who accept the neo-Freudian point of view, it contains a hard kernel of truth.

The new openness to sexual matters in communal groups ranges from the merely verbal in sensitivity sessions with "moderate" groups, to group marriage as practised by groups such as the "Expanded Family" in New York, and finally to perhaps the ultimate in sexual permissiveness as epitomized by the Sexual Freedom League with headquarters in New York.

The openness toward sex of some of the new community-building young is in marked contrast to a great majority of the nineteenth-century American utopists who severely regulated sexual activity. Yet both the new and the old utopians,[13] by rejecting bourgeois standards concerning sex, set themselves radically apart from the mainstream of society.

Corporate Capitalism—Technology. It has long been held by anarchist and Marxist scholars that capitalism uproots man from his community, alienates him from his labor and in essence turns him into an object[14]—a cog in the industrial wheel. The traditional answer to this problem—a form of democratic or revolutionary socialism—has held little appeal to the twenty-year-old communal utopists of today. They feel that the eastern bloc countries are simply another establishment, and further that the socialist nations have accepted the standards and goals of capitalism—that is the need for increasing technological sophistication, production, and consumption.

"It doesn't take a weatherman," as the classic Bob Dylan song points out, "to know which way the wind blows." The winds of technological affluence in the United States have not promoted a "happy consciousness" among the dissident young. While millions of poor whites, blacks,

13. See Chapter Three, "Early Communitarianism in America."
14. The Southern African intellectual, Cesaire, describes this process as "chosification," (turning men into things).

Chicanos, and Indians were struggling to share in the minimal standard of health and comfort, more affluent young people were consciously lowering their living standards to escape (albeit temporarily in some cases) what they considered to be the tyranny of the machine.

Mexican-Americans near Taos, New Mexico have greeted the arrival of primitive hip communes with a mixture of astonishment and antipathy. How could anyone consciously reject "the good life"—the very thing they were fighting for?

Perhaps part of the reason the hip can reject the "good (comfortable) life" is that they have become aware that the price for enhancing one's "standard of living" is frequently paid in damage to the human spirit. It is one thing for a man to struggle to feed his family but quite a different matter for a young student born in relative affluence to "bust a gut" (literally in that he may develop ulcers) for a $60,000 home rather than a $35,000 model.

Theodore Roszak in his *The Making of a Counter Culture*[15] maintains that many of the young reject the rational-technological society because they have found its promises to be false and disastrously so. Thus, the students "go back to the land," back to the "tribal" existence, and back to a magical or shamanistic view of the earth. ". . . It is nothing new," Roszak argues, "that there should exist antirationalist elements in our midst. What is new is that a radical rejection of science and technological values should appear so close to the center of our society, rather than on the negligible margins. It is the middle-class young who are conducting this politics of consciousness, and they are doing it boisterously, persistently, and aggressively—to the extent that they are invading the technocracy's citadels of academic learning and bidding fair to take them over." [16]

15. Garden City, New York: Anchor Books, 1968.
16. *Ibid.*, p. 51.

2

UTOPIA AND COMMUNE

"Gluttony, greed, lack of compassion have caused America to become the most despised nation on Earth. The sad thing is, my Polish Lady tells me, we were throughout her youth, and still are, or could be, the hope of all.

We face great holocausts, terrible catastrophies, all American cities burned from within, and without.

However, our beautiful planet will germinate—underneath this thin skin of city, Green will come on to crack our sidewalks! Stinking air will blow away at last! The bays flow clean! . . . In the meantime, stay healthy. There are hundreds of miles to walk, and lots of work to be done. Keep your mind. We will need it. Stake out a retreat. Learn berries and nuts and fruits and small animals and all the plants. Learn water."

☺ *The Digger Papers, 1968*

What is utopia? Further, what is communal living? The answers we give to these questions depend to a large extent on our own biases. The word "utopia" is taken from Thomas More's work of the same name. It is literally translated as "nowhere." It is true that if we look for a perfected society on earth it is indeed "nowhere." Men's dreams are not often limited by the "possible," however; and dreamers continue to plan for heaven on earth. To an orthodox Marxist, utopian dreamers were wrong precisely because their dreams were without conflict. Henri de Saint Simon, Charles Fourier, and Robert Owen, three early European utopists, published tracts and books in the second and third decades of the nineteenth century in which they constructed ideal societies. All three were later soundly condemned by Marxists for their "naive" belief that ideal societies could be created by moral persuasion. Hence, the word "utopian" as used by Marx and Engels was a term of derision. Marxist radicals did agree with the earlier utopian thinkers that a new social organization must be created to promote a peaceful, nonalienated, and just society. They were sure, however, that the larger bourgeois society must fall before the new society and "new man" could exist.

If those seeking a perfect, conflict-free society were seen as naive and historically irrelevant by the socialist revolutionaries, they were regarded as a positive evil by the church. Orthodox Christian doctrine held that man's nature was evil since he was "born in sin." Therefore, the attempt to seek the perfect society on earth was an illusion as well as a gross sin.[1] Of course, it was also true that the church had seen the dynamics of an earlier mass-based movement to build heaven on earth—the Anabaptists. A good deal of blood was shed before this utopian movement of the sixteenth century Europe was destroyed.[2]

The essence of early utopian thought, then, was a dissent against the *status quo*. The quest for perfection, by definition, means a rejection of the present. More than this, there is in most early utopian thought the idea that man has potential for goodness and that he can reach that goodness if he is placed in the proper kind of society.

In 1817, Robert Owen, a classic Utopian thinker, challenged the idea that the poor were innately lazy, ignorant, or morally inferior. "Either give the poor a rational and useful training," he said,

> or mock not their ignorance, their poverty, and their misery, by merely instructing them to be conscious of the extent of the degradation under which they exist. And, therefore . . . either keep the poor, if you now can, in the state of the most abject ignorance, as near as possible to animal life, or at once determine to form them into rational beings, into useful and effective members of the state.[3]

1. For a modern example of this viewpoint see Thomas Molnar, *Utopia: The Perennial Heresy* (New York: Sheed and Ward, 1967).

2. Norman Cohen's excellent historical work, *The Pursuit of the Millennium* (New York: Harper and Row, 1961) holds that the medieval Anabaptist movement became the prototype for almost all European upheavals and revolutions in the nineteenth and twentieth centuries.

3. Quoted in Rowland Hill Harvey, *Robert Owen: Social Idealist* (Berkeley: University of California Press, 1949), p. 35.

Frank Manuel[4] makes a useful and enlightening classification of Utopian varieties. The earliest he calls "Utopias of calm felicity." These Utopias are typified by Thomas More's sixteenth-century *Utopia* or Francis Bacon's *New Atlantis.* These pre-French revolutionary planners concerned themselves with equality, moderation, and the establishment of a tranquil environment.

A second variety outlined by Manuel is the "open-ended utopia of the nineteenth century." Unlike the utopia of calm felicity, this ideal society is hierarchical and dynamic. The Saint Simonians and Fourierists, for example, felt that the new society would utilize science to provide an unequal but just system of rewards for everyone.

Finally Manuel describes what he calls "contemporary eupsychias." This latter type (as epitomized by Teilhard de Chardin, Julian Huxley, Eric Fromm, and Abraham Maslow) is evolutionary, universalistic, and deals with potentiality for ever-increasing self-awareness in man.

CHARACTERISTICS OF COMMUNAL SOCIETIES

If utopian schemes all involve a rejection of current society and an acceptance of social experimentation, how then do they differ from communal living? In one sense, "communes"[5] may be regarded as a subclass of the larger category of Utopias. A modern "commune," as the word is used here, differs from other utopian forms in three basic ways. First, *communal societies specifically reject the concert of hierarchy or gradations of social status as necessary to the social order.* Not all proposed Utopias are egalitarian. If one were to go back to what is perhaps the earliest planned society, Plato's *Republic,* he would find a society more rigorously stratified than that of current America. As mentioned above, neither Saint Simon or Fourier sanctioned pure communism. In fact, Saint Simon, for all his radical rejection of religion and the aristocracy, proposed that the leadership of his new society be composed of benevolent scientists, technicians, businessmen, and bankers.[6]

4. Frank Manuel, "Toward a Psychological History of Utopias," in *Utopias and Utopian Thought* (Boston: Beacon Press, 1967), pp. 69–95.

5. Communes in this sense would not encompass monastic communities, for while they have traditionally rejected the larger society, they are hierarchical and lacking in social experimentation.

6. Henri de Saint Simon, *Social Organization: The Science of Man* (New York: Harper and Row, 1964).

A good argument can be made for the idea that unequal access to power does not promote maximum understanding. For example, it is axiomatic in the study of race relations that contact between unequals (*e.g.* sharecroppers and plantation owners) does little to diminish prejudice. It is, in point of truth, more likely to heighten suspicion and hostility. On the other hand, noncompetitive contacts between Blacks and whites in an egalitarian setting (*e.g.* as students together in college) tends to diminish prejudice. Ronald Sampson in his *Psychology of Power* states the case in this way: "To the extent that we develop our capacity for power we weaken our capacity for love; and conversely, to the extent that we grow in our ability to love we disqualify ourselves for success in the competition for power."[7] This, then, is the position of many communalists currently in the United States and Europe.

Hegel master-slave

Unfortunately, the ideal of a nonhierarchical society is conceptualized far easier than carried out. One anarchist group formed a communal farm in upstate New York in the late 1960s and found a legion of difficulties in promoting an egalitarian social arrangement.

"I suppose," one of the founders says,

> our first and worst economic argument was whether or not to buy chickens. At first it was incredible how little a problem money had been. Whoever came just threw in whatever they had—$100 or $200 perhaps—and we'd live off that until someone got a tax return, a welfare check or whatever. We never did spend more than $24 a week on food—even when there were 30 people. But the chicken crisis involved all sorts of things. Did we need eggs? . . . who would build the chicken coop? This was the first time I remember hearing anyone say, "Well, *I* won't give money for chickens."—using money as a weapon, a personal source of power. And it wasn't long before money again became a personal possession.[8]

This statement reveals both the idealism and frustration intrinsic to the attempt by communalists to produce a classless society. Individuals socialized in an essentially competitive, money-oriented society are not easily disposed to change their attitudes and commitments to private property. This is not to say that these changes in feeling cannot come about. The Israeli kibbutz provides rather strong evidence to the effect

7. Ronald V. Sampson, *The Psychology of Power* (New York: Random House, 1968), pp. 1–2.
8. Joyce Gardner, "Cold Mountain Farm," *The Modern Utopian* 2, no. 6, p. 4.

that (given certain circumstances) a "classless" [9] society can be achieved on a small scale.

Another belief tends to set modern communalists off from other Utopists. That is the fact that *modern communal societies all maintain that the scale of society as currently organized is too large.* Unlike Marxist utopians who conceive of an entire world community, communalists tend toward the anarchistic idea that man can relate in a meaningful way with a limited number of like-minded individuals.

Of course, arguments over the size of the ideal community are as old as society itself. Aristotle found the population limitation for the ideal *polis* to be the number which could be conveniently addressed by a single orator. Charles Fourier limited his utopian "phalansteries" to about sixteen hundred persons. Moreover, sociologists from Durkheim to the present have argued that rapid increases in population density in limited areas may provide dynamic social change, not all of which is healthy. Current ecologists also tell us that population increase is perhaps the root cause of the destruction of man's "living space." [10]

Beyond this however is the simple notion expressed by many communalists that individuals who participate in modern, large-scale institutions are less able to relate to others in a truly "human" way. Hence, it is no accident that American universities experiencing the most convulsive student unrest in the 1960s were the largest (and perhaps most bureaucratic) in the nation.

What kind of limits do modern communalists place upon their newly formed social orders? The answer to this question could be given in either a real or ideal sense. One current communalist, writing in *The Modern Utopian* in 1968, structures his ideal community in this way.

> [It] is built of several . . . family units, along with additional individuals and couples not so firmly attached to a subgroup or in process of joining; total size or planned maximum usually is 50 or 60, rarely exceeding 100. 60, incidently, is the size of the psychological "primary group"—the size where every member can know every other's first name." [11]

9. The abolition of social classes does not necessarily involve the lack of prestige or "ranking" in a kibbutz. It merely involves an obliteration of the division of labor and the traditional rewards related to occupational roles. See Melford E. Spiro, *Kibbutz: Venture in Utopia* (New York: Schocken Books, 1963).

10. See Paul R. Ehrlich, "Too Many People" in Garrett de Bell's *The Environmental Handbook* (New York: Ballantine Books, 1970).

11. C. P. Herrick, "Designing for Community," *The Modern Utopian* 2, no. 6, p. 11.

The fact is, however, that many current communal groups are a good deal less sure of the optimum size than the last quotation indicates. Some struggling groups are forced to advertise for new members simply to maintain themselves while other communities, such as "Drop City" near Boulder, Colorado, have made it known that new residents are not needed nor are tourists and visitors welcomed. In actuality, most communal groups in America today, whether hippy, digger, or religious in nature, are so handicapped economically or by lack of space that potential members must be judged by ability to make financial contributions as well as by personal characteristics.

One final characteristic which sets modern communal movements apart from other utopian ventures is that *communal societies are consciously antibureaucratic in structure.*

The great literary anti-Utopias of the twentieth century, *Brave New World*[12] and *1984*,[13] were self-consciously bureaucratic nightmares. Their total planning and manipulation of the individual had been given a good deal of reality by the advent of two totalitarian bureaucratic states, Stalinist Russia and Nazi Germany. It is rather a moot question as to whether Soviet Russia or America is more bureaucratic currently. The fact is that both corporate capitalism and what Eric Fromm calls the "state capitalism" of the U.S.S.R. are highly bureaucratic. It may be argued that bureaucracies are not by nature inherently antidemocratic, but few, I think, would argue that membership in a bureaucracy is very satisfying emotionally.[14] In fact, Kenneth Keniston, a Yale psychiatrist, attributes much of the turmoil of adolescence as a rebellion against the dull, repetitious future occupation awaiting him in a large scale bureaucracy. According to Keniston,

> the young person abandons a world of directness, immediacy, diversity, wholeness, integral fantasy, and spontaneity. He gains abstraction, distance, specialization, monotony, dissociated fantasy, and conformity. Faced with [this] . . . transition . . . the youth can only hesitate on its threshold. . . . The humanization of childhood has been accompanied by a dehumanization of adulthood.[15]

12. Aldous Huxley, *Brave New World* (New York: Harper & Row, 1946).
13. George Orwell, *Nineteen Eighty-Four* (New York: New American Library, 1960).
14. Peter Blau takes this position in his *Bureaucracy in Modern Society* (New York: Random House, 1956).
15. Kenneth Keniston, "Alienation and the Decline of Utopia," *The American Scholar* 29 (1961): 172.

All of us have resented at one time or another the emotional detachment involved in bureaucratic dealings. Many of us, too, would like very much to recapture the nonblasé thrills of childhood. In order to rid ourselves of the stifling bureaucratic "demon," however, a price must be paid. Basically, the cost of removing bureaucracy from this or any other society is that of lowering one's standards of living. The basic difference between communalists and "straight" (middle-class) individuals is that while both may react negatively to bureaucracy, middle class individuals are not prone to lower their own level of living to do away with bureaucracy. Many drop-outs from the larger society have (in part) solved their financial problems by panhandling or the like. For a communal group to free itself from the bureaucracy of the larger society, it must first become self-sufficient economically. For this reason, many communal "drops-outs" seek a simpler agrarian means of subsistence. Current publications, such as the *Whole Earth Catalog*[16] give modern communalists access to information concerning "organic gardening" (*i.e.*, gardening without the use of synthetic chemicals), herbs, construction of homes with few tools, natural childbirth, and the like. In short, it is a kind of source book for the new communal self-sufficiency.

One interesting sidelight to the problem of avoiding the bureaucratic ethos is the fact that the Haight-Ashbury district of San Francisco, the birthplace of hippy communal living, became so commercialized (bureaucratized) that many hippies left the area. On October 6, 1968, the residents of the area held a funeral procession to dramatize their feeling that Haight-Ashbury was dead as a hip community.[17]

WHAT TO LOOK FOR IN MODERN UTOPIAS AND COMMUNES

The Motivation of Membership. C. Wright Mills, in one of his most perceptive essays,[18] maintained that the central task of politics was to turn "the personal troubles of the milieu" into the public issues of social structure. What Mills meant was that many of the "personal" problems of the individual, such as alienation or anxiety, have their roots in the inadequate functioning of social institutions. Political activists who

16. Menlo Park, California: Portola Institute, 1969.
17. *Berkeley Barb*, October 12, 1968.
18. C. Wright Mills, *The Sociological Imagination* (New York: Oxford University Press, 1959).

"make the communal scene" believe they have located the roots of their personal troubles in the social structure.

"We're unhappy because of the war, and because of poverty and the hopelessness of politics, but also because we sometimes get put down and alone and lost." [19] James Kunen's analysis of his motivations for participation in the Columbia University confrontations of 1968 seems both honest and even a little whimsical. The ugly war in Vietnam, racism, and poverty are the causes of much unhappiness for the politically aware young. No motive is pure, however, in the sense that it deals only with political or social issues. It is possible that some young advocates of social change may be no more aware of their true motives for action than the conservative is of his motivation for holding to the *status quo*.

Even though communalists are not always aware of their motivations, many are able to articulate their reasons for community building in a rather delightful way. Art Rosenblum, a mystical communalist living in Pennsylvania, describe his desires in this way.

> We want tremendous freedom; we want to be so free that we could just flap our arms and fly through the air. . . . We want to be able to explore all the worlds, and have tremendous adventures. We want those adventures to be meaningful also—not just excitement for its own sake. We want to be understood—fully comprehended by others, and to comprehend them fully and not only people but also all the rest of creation. We want to be free to be fully open and honest about everything with everyone. We want to be completely free of guilt and shame so that we could even be naked in the presence of others and find it natural.
>
> Though we long for adventure, we also long for security; the absolute security of immortality, and of never having to grow old and feeble. We also long to have children and bring them up in an environment where they will be loved and cared for by all.[20]

The Constituency. Every social movement must appeal to a segment or segments of society which are potentially amenable to the movement's ideology. The potential constituency of the modern communal-

19. James Simon Kunen, *The Strawberry Statement: Notes of a College Revolutionary* (New York: Avon Books, 1970), p. 11.

20. Art Rosenblum, *Aquarian Age or Civil War* (Philadelphia: Aquarian Research Foundation), p. 10.

utopian movement is the young, disaffected, post-high-school, college and postcollege individual. In contrast to working-class youths, many middle-class college students speak of their resentment of the "plastic" or "inauthentic" nature of bourgeois occupations, politics, religion, and respectibility. It is no challenge for many upper-middle-class youth to find a good job or economic security. The status-seeking rat race has been debunked for them in many popular and semipopular sociological studies. Without a doubt, a number of the affluent young are psychologically ready to experiment socially. Since their lives have not prepared them effectively for the sometimes primitive living of communal environs, their adjustment could not be expected to be easy, and one would expect their rate of disaffection to be rather high.

Constitutive Ideas. This term, used by Rudolf Heberle in his *Social Movements*[21] to refer to the *raison d'etre* for a movement, encompasses as well the movement's conception of its final goals and the means of attaining those goals. The constitutive ideas of many utopian and communal ventures are necessarily vague since experimentation and fear of dogmatism are common to members.

The *Directory of Intentional Communities* (1968)[22] is of some aid, however, in roughly demarking the ideologically important elements in the thought of current utopian and communal groups. Two hundred and twelve groups are listed in the *Directory*, with about half of this number stating a few descriptive words about their purposes or activities. Twenty-two groups described themselves as "christian" while four other groups stated they were organized on nonchristian religious principles, such as Yogi, Buddhism, or reincarnation. (Three groups claimed to have created new "independent" religions.)

Some communal organizations claimed that a reorganization of sex mores was necessary to the ideal society. Eight groups of this variety cited a form of group marriage as an organizing principle for them. Two other groups proposed the reorganization of society in terms of "free love."

A number of modern communal organizations are politically concerned. Five groups in the *Directory* categorize themselves "revolutionary," while a slightly larger number (seven) refer to themselves as anarchists.

21. Rudolf Heberle, *Social Movements* (New York: Appleton-Century-Crofts, 1951).
22. Published by the Alternatives Foundation, Berkeley, California.

It must be remembered that not all communalists wish to simply escape from the larger world. In fact, fifteen communal organizations described themselves as "service oriented," that is to say that they worked with the disadvantaged, the aged, the emotionally disturbed, or through their own educational institutions.

Other constitutive ideas for the new Utopians are economic at base. Nine groups were in the process of establishing cooperatives of land or housing, while four stressed going back to a subsistence-level economy with organic or "microbiotic" gardening.

Finally, a small number of groups (three) admitted to following psychologist B. F. Skinner's planned utopia, Walden Two.[23] A few others mentioned the establishment of an art colony and a permanent sensitivity group.

At this point it is easy to see the nearly total diversity between these utopian and communal groups. One should not forget, nevertheless, that all these groups have opted for a rejection, to some degree, of the larger society, its social forms, and its system of rewards.

Group Maintenance. How do modern communal and utopian groups maintain themselves as ongoing systems? Since the rebirth of the communal phenomena is so new, the question as to how communal groups maintain themselves is still open. The problems these groups face, however, are no different in kind than those of the maintenance of a family or nation. They consist quite simply of insuring control over certain resources (food, shelter, and the like) and the cooperation of the individuals within the group.

Both problems are in fact complex. Communal groups that "go back to the land" sometimes find hostile neighbors or perhaps a hostile climatic or physical environment. Hip communes near Taos, New Mexico, have found both, and violence has sometimes erupted between local "cowboys" and hip residents. Another hip commune—"Oz," near Meadville, Pennsylvania—was, in essence, destroyed by the intolerance of the local residents who had complained of the "unsanitary living" conditions and "immorality" of the group. When one resident of the commune was taken to the hospital with hepatitis, a local woman protested the fact that gamma globulin was given them. "I would let them live there and die." [24] Finally "an injunction was nailed to the front of

23. B. F. Skinner, *Walden Two* (New York: Macmillan, 1962).

24. Robert Houriet, "Life and Death of a Commune Called OZ," *New York Times Magazine*, September 15, 1969, p. 100.

the farmhouse forbidding the use of the premise for fornication, assignation, and lewdness." [25] Local stores refused to sell supplies to the communalists. Finally, in September of 1968 the group scattered to avoid the legal and extralegal harassment they confronted.

Another problem of an internal nature will face communal groups that plan for permanence. (Many, of course, do not.) This is the problem of "socializing" the younger members of the group to accept the basic values of the community. Twin Oaks, a *Walden Two* community in Virginia, accepts the idea that children should be disciplined but never punished in a physical sense. This is consistent with the group's goals of organizing a community on behaviorist lines.

Nevertheless, Mannheim,[26] Heberle,[27] and others have argued that the problem of "political generations" is a real one. Communitarian goals may be avidly sought by one generation and rejected by the next. Individuals who are in fact able to carve out a utopia for themselves sometimes find they have created nothing less than a prison for their sons and daughters.

It has been said that those ignorant of the past are doomed to repeat its errors. For that reason a review of the rise and fall of earlier American utopias and planned communities may be helpful to understand the stresses and strains of current community-building.

25. *Ibid.*
26. Karl Mannheim, *Essays on the Sociology of Knowledge* (London: Routledge and Kegan Paul Ltd., 1952).
27. Heberle, *op. cit.*, pp. 118–27.

3

EARLY COMMUNITARIANISM IN AMERICA

A map of the United States showing the existence of utopist communities extant during the eighteenth and nineteenth centuries would point to the pervasive nature of communal life in early American society. John Humphrey Noyes, for example, in investigating the Owenite and Fourierist colonies in existence from 1824 to 1850 cites four Owenite groups in Indiana, three in New York, two in Ohio and Pennsylvania, and one in Tennessee.[1] Fourierist groups were found in this number: eight in Ohio, six in New York, six in Pennsylvania, three in Massachusetts, three in Illinois, two in New Jersey, two in Wisconsin, two in Indiana, and one in Iowa. The number of individuals living in these communal arrangements at this period Noyes estimates at about 8,640. The amount of land reported is over 135,000 acres. Again, it must be reiterated that this number does not include the numerous Mennonite groups, Spiritualist associations, Moravians, Zoarites, Icarians, Shakers, and Latter Day Saints. It is, in reality, impossible to account for all the groups in the New World who sought heaven on earth. One can be certain that the image many Europeans had of America as a land innocent and unspoiled added to its attraction as a seedbed for community-building.

SHAKER SOCIETIES

One of the earliest and most notable of the American utopist movements was the United Society of Believers in Christ's Second Appearing —usually known as the Shakers. Ann Lee, its founder and prophetess, was born in Manchester, England, in 1736. She was given to visions as

1. John Humphrey Noyes, *History of American Socialisms* (New York: Dover Publications, Inc., 1966).

were many religious nonconformists of her day[2] and was frequently imprisoned for her heretical views. While in prison in 1770 she was granted a vision of Christ which showed her

> a full and clear view of the mystery of iniquity, of the root and foundation of human depravity, and of the very act of transgression committed by the first man and woman in the Garden of Eden. [the sexual act itself][3]

Mother Ann, as she was now called, left England in 1774 for New York. Her husband William was said to have revolted somewhat from her newly imposed rule of celibacy; at any rate he soon eloped with another woman. Mother Ann's movement was clearly millenarianism. A typical prophetic utterance by the "Shaking Quakers" or "Shakers" as they were often called is as follows:

> . . . Amend your lives. Repent. For the Kingdom of God is at hand. The new heaven and new earth prophesied of old is about to come. The marriage of the lamb, the first resurrection, the new Jerusalem descending from above, these are even now at the door. And when Christ appears again, and the true church rises in full and transcendant glory, then all anti-Christian denominations—the priests, the church, the pope—will all be swept away.[4]

The religious revivals sweeping the burned-over districts of New York were not without their effect on the Shakers. Shakerism became more ritualistic in the new world and from all accounts, even more dionysian in technique as well.

> In the best part of their worship everyone acts for himself, and almost everyone different from the other: one will stand with his arms extended, acting over odd postures, which they call signs; another will be dancing, and sometimes hopping on one leg about the floor . . . ; another will be prostrate on the floor . . . some groaning most dismally; some trembling extremely; others acting as though all their nerves were convulsed; others swinging their

2. Many of her religious ideas were taken from the "Camisards," a French Protestant group which had fled to England to protect itself from religious persecution.

3. Quoted in *Heavens on Earth* by Mark Holloway (New York: Turnstile Press, 1951) p. 57.

4. Edward Deming Andrews, *The People Called Shakers* (New York: Dover Publications, Inc., 1963), p. 6.

> arms with all vigor . . . They have several such exercises in a day, especially on the Sabbath.[5]

The Shakers, like the early Christians, were convinced that the return of Christ would take place in their generation. Their task then was to withdraw from the world and to live in harmony until the cosmic event occurred. This was to be coupled with an intensive missionary effort through the eastern states. Between 1781 and 1783 the Shakers' campaign gained them many converts and much persecution as well.[6] Mother Ann died in 1784 and was succeeded by Joseph Meacham or "Father Joseph" as he was called.

Father Joseph began to establish communities on the basis of a common charter. He also drew up the code and principles which were to govern the first eleven societies. The covenant stated that

> All members might have an equal right and privilege, according to their calling and needs, in things spiritual and temporal. And in which we have a greater privilege and opportunity, of doing good to each other, and the rest of mankind and receiving according to our needs, jointly and equally, one with another, in one joint union and interest.[7]

Communism was a spiritual answer to the evils of unrestrained egoism, just as celibacy was the answer to unrestrained lust. All were to dress in plain and modest apparel; all were to live in a common dwelling and in "consecrated labor."

> Hand labor was an integral part of the process if only to avoid poverty. . . . But manual work was glorified from higher motives. It was good for both the individual soul and the collective welfare, mortifying lust, teaching humility, creating order and convenience, supplying a surplus for charity supporting the structure of fraternity, protecting it from the world, and strengthening it for increasing service.[8]

The structure of the movement had as its basic unit "the family" which was composed of "brother" and "sister" celibates living in the same buildings. "Brothers" "sisters" were not allowed to pass each

5. *Ibid.*, p. 28.
6. This was due in part to their refusal to bear arms in the Revolutionary War.
7. Andrews, *op. cit.*, p. 62.
8. *Ibid.*, p. 104.

other on the stairs, speak to each other alone, eat together or the like. Usually about four families constituted a "society" which might contain as many as eighty persons.

> Each family was administered by two elders and two elderesses, who formed the ministry responsible for both the temporal and spiritual welfare of the societies. . . . The authority of this self-perpetuating ministry was absolute, without appeal. It appointed its own successors, without election and exacted implicit obedience. The leading elders . . . heard all confessions, knew the whereabouts and occupations of every shaker in their family, conducted the initiation of novices, controlled the movements of trustees in their dealings with the world, and exercised their power in numerous other ways.[9]

By 1830 the Shaker Church had reached a membership of about 5,000, in about eighteen large societies. Nordhoff's census of the Shaker communities in 1874 found less than half that number (2,415) active. The last community to fall, in New Lebanon, New York, was founded in 1787 and lasted until the last family moved in 1947.[10]

What were the causes of the eventual demise of the order? According to Andrews, economic problems were crucial.

> Economic factors, which played an important role in the success of the movement were involved in its decline. Outstanding were (1) the expense of maintaining the order, (2) the Shaker policy regarding land ownership, (3) the mismanagement which followed in the wake of prosperity, and (4) the impact of the industrial age.[11]

Further, the enforced celibacy of the group precluded their natural increase. By 1840 an ideological split weakened the organization also, as a "liberal versus conservative" schism threatened. Elder Frederick Evans of New Lebanon represented the liberal forces, who advocated closer relations with the world and active social reform. He corresponded with Henry George on the single tax and with Tolstoy on cooperation and nonresistance. He was, of course, violently attacked by traditionalists

9. Holloway, *op. cit.*, p. 68.
10. Charles Nordhoff, *The Communistic Societies of the United States* (New York: Dover Publications, Inc., 1966) p. 256.
11. Andrews, *op. cit.*, p. 226.

who feared the contamination of the world. This conflict could only injure the dynamics of the societies' "outreach" to the world.

In retrospect, the Shaker community was much more successful than many of its later utopian counterparts. John Humphrey Noyes credits the Shakers with far-reaching influence.

> France had also heard of Shakerism before Saint Simon or Fourier began to meditate and write Socialism. These men were nearly contemporaneous with Owen, and all three evidently obeyed a common impulse. . . . It is very doubtful whether Owenism or Fourierism would have ever existed, or if they had, whether they would have ever moved the practical American nation if the facts of Shakerism had not existed before them and had gone along with them.[12]

GERMANIC UTOPISTS IN THE NEW WORLD

Rappites. The results of the Anabaptist and of other schismatic Protestant movements in Germany became apparent in the American communal scene in the eighteenth and nineteenth centuries. Many Germanic separatists came to the North American continent with a rigidly ascetic desire to build the good society.

One of the earliest movements of this variety was led by George Rapp, who sailed from Wurttemburg, Germany, in 1803 for Pennsylvania. He bought 5,000 acres of virgin land north of Pittsburgh the next year and immediately sent for his followers, who numbered about 750.[13]

Like other pietistic sects, the Rappites believed in a "return to the Bible" and a retreat from an incurably evil world. They adopted celibacy soon after their voyage to the New World and "discovering that they included among their number many who were too old, too infirm, or too poor to be able to maintain themselves, the Society resolved to adopt communism." [14]

Their community of Harmony in Pennsylvania was relatively prosperous but lacked water and was unsuitable for vine-growing. In 1814 the group bought 30,000 acres in the Wabash Valley of Indiana and sold

12. Noyes, *op. cit.*, p. 670.

13. Holloway, *op. cit.*, p. 89. See also Phebe E. Earle, *Pennsylvania Dutch and Other Essays* (Philadelphia: Lippincott, 1874).

14. *Ibid.*, p. 90.

their original settlement. Since their neighbors were stricken with malaria, they sold their New Harmony to Robert Owen for $150,000 and moved back to Pennsylvania to found the new community of Economy.

In 1875 Nordhoff found, "the society . . . reported to be worth from two to three millions of dollars." [15] Because the community had been held together by the charisma and executive abilities of Father Rapp, it declined severely with his death. Nordhoff comments that:

> They hold that the coming of Christ and the renovation of the world are near at hand. . . . Father Rapp firmly believed that he would live to see the wished for reappearance of Christ in the heavens. So vivid was this belief in him that it led some of his followers to fondly fancy that Father Rapp would not die before Christ's coming; and there is a touching story of the old man, that when he felt death upon him, at the age of ninety, he said, "If I did not know that the dear Lord meant I should present you all to him, I should think my last moments come." [16]

The Rappites had passed the zenith of their power at Nordhoff's visit to them and had been recently shaken by schisms. When asked how they expected to carry on another decade by Nordhoff, the aged leadership replied, "The Lord will show us a way. . . . We have not trusted him in vain so far. We trust him still. He will give us a sign." [17] The community was to disintegrate by the beginnings of the twentieth century.

Inspirationists—The Amana Colonies. A later migration to America was made by another group of German pietists, the Inspirationists. Christian Metz, a carpenter, founded several cooperative settlements near Herrnhaag, Arnsburg and Marienborn in Germany.[18] Life in Germany became impossible for the sect, however, due to the group's refusal to take oaths or submit to military service. On October 26, 1842, Metz and companions landed in New York. They bought about 5,000 acres of land near Buffalo but later traded this for 20,000 acres of excellent land in eastern Iowa. The Inspirationists, unlike the Shakers or Rappites, were urban craftsman. According to Holloway,

15. Nordhoff, *op. cit.*, p. 93.
16. *Ibid.*, p. 86.
17. *Ibid.*, p. 95.
18. Holloway, *op. cit.*, p. 169.

They found that communism, which had not been envisaged when they left Germany, was the only practical means of providing industrial, as opposed to agricultural employment for those members who were artisans; and they were once again commanded by inspiration to adopt a course of action that was essential to their survival.[19]

Some of the "Rules for Daily Life" printed by those living in the Amana (Iowa) Colonies were as follows:

> To obey without reasoning God and through God our superiors; to abandon self, with all its desires, knowledge, and power; [and to] have no intercourse with worldly minded men; never seek their society; speak little with them, and never without need; and then not without fear and trembling.[20]

Women in the colonies were subjugated to a large extent. Like the Shakers, the Inspirationists restricted social contacts between the sexes. Formal education was seen as leading first to worldly concerns and, finally, to heresy. At least once in every year an *untersuchung* or inquisition of the community was held. Each member was expected to make a confession of all his sins, transgressions, and shortcomings to the "inspired person" or elder.

The inevitable confrontation with the modern world was to await the separatists. By the turn of the nineteenth century the full scale communism of the group had caused disaffection by the young. In order to adjust to the new circumstances, the group in 1932 formed a joint stock cooperative society which became even more successful in an economic sense.[21] Education became less the mark of heresy, and by the middle of the twentieth century, many young men were attending college at their own expense. The population of the colonies remained stable at about 1,400.

Mennonites. The German Protestant group having the most direct connection to the Anabaptists was the Mennonite or Evangelical Anabaptists. Menno Simmons, a Dutch Roman Catholic priest, left his orthodox faith in 1540 for the Anabaptist sect. He was driven out of the Netherlands for his beliefs and settled in Northwest Germany.

19. *Ibid.*, p. 170.
20. Nordhoff, *op. cit.*, p. 51.
21. Holloway, *op. cit.*, p. 219.

Simmons drew great followings with him because of his teachings. His missionary work took him to Poland, Flanders, and West Prussia. By the time of his death, his followers, the Mennonites, had become a powerful utopist force in Northwestern Europe. They were constantly at odds with secular authorities in Prussia because of their pacifism and because in 1787 they were required to pay tithes to the established Lutheran Church.[22]

In 1788 the Russian minister offered to grant special privileges and concessions to the Mennonites to induce them to move their communities to his country. Ten years later 18,000 believers formed colonies on the Volga.[23] Village communes were formed and the Mennonite societies became economically well-balanced and prosperous. Once again, however, the demand that they bear arms set the pietists looking for new lands that would respect their nonviolent attitudes. The elders of the group negotiated unsuccessfully with the United States, but successfully with the Canadian government and arrived in the latter country in 1873. The migration involved 1,336 families, with an estimated 7,500 members, who settled for the most part in Manitoba.[24]

Since the Mennonites had done a great deal of colonizing in the past, they had little trouble in adapting themselves to their new environs

> The settlement pattern of the Mennonite village was that of the northeast German colonial *Gewanndorf* characterized by a combination of line village with open-field economy. . . . The open-field system is closely associated with the practice of crop rotation. . . . The village organization . . . may be called the solidaristic type of settlement for it pre-supposes and fosters strong coherence, intensive interaction on a face-to-face level, readiness to cooperate and offer mutual aid . . . and which is enforced by strict social control based on both inner and external sanctions.[25]

The Mennonite community based on cooperative agriculture was one of the most successful utopian ventures on the American continent. It had survived in various forms and in a variety of areas for over 250 years. Yet it was not totally immune to social change that confronted it. By the 1920s certain signs of deterioration within the community be-

22. E. K. Francis, *In Search of Utopia: The Mennonites in Manitoba* (Altona, Manitoba: O. W. Friesen & Sons Ltd., 1955).
23. *Ibid.*, p. 19.
24. *Ibid.*, p. 50.
25. *Ibid.*, p. 64.

came evident. Some of the young, resentful of their restricted lives, left the faith. Moreover, theological schisms split the organization. By the 1940s several groups had abandoned the last vestiges of the old Mennonite ritual. About 45,000 descendants of the original community lived in midwestern Canada in the 1950s. As E. K. Francis states,

> Utopia is farther beyond the horizon than ever. But Manitoba's Mennonites have found social and psychological security in their well organized communities, and sufficient wealth to give them a sense of satisfaction and contentedness.[26]

OWENISM IN AMERICA—THE NEW HARMONY EXPERIENCE

Robert Owen came to the United States in 1824. "I am come to this country," he said a year later, "to introduce an entirely new system of society; to change it from an ignorant, selfish system to an enlightened social system which shall gradually unite all individuals into one and remove all causes for contest between individuals." [27]

Owen had convinced himself and several others that the way in which man was to be changed was through a change, initially, in man's physical environs. An ideal community began with new buildings. They were to be shaped in a hollow square 1,000 feet long and would contain lecture halls, a school, kitchens, and apartments.

John Humphrey Noyes[28] lists eleven societies of the "Owen epoch" founded in 1826. They included Forrestville Community in Indiana, with sixty members; Haverstraw community, New York, with eighty members; Kendal community, Ohio, with two hundred members; Yellow Springs in Ohio, with four hundred members; and Nashoba, Tennessee, with only fifteen members. Most failed within a period of less than three years.

Owen's greatest challenge at community-building came, however, with the founding of New Harmony Community. He had contacted Father Rapp for the sale of the Indiana community in 1825. His son, Dale, reported that by January of the next year

> my father must have been as well pleased with the condition of things at New Harmony, on his arrival, as I myself was. At all

26. *Ibid.*, p. 278.
27. Holloway, *op. cit.*, p. 104.
28. Noyes, *op. cit.*, p. 15.

events, some three weeks afterwards, he disclosed to me his intention to propose to the Harmonites that they should at once form themselves into a Community of Equality, based on the principle of common property.[29]

The community was, in fact, to be called the "New Harmony Community of Equality." Owen assumed executive control of the group for one year to alleviate possible dissension. One factor which did exacerbate other problems in the community was Owen's liberal position on religion. New Harmony rapidly gained the reputation of being a hotbed of atheism. By the end of the first year, nonetheless, the community had gained 1,000 members. Moreover, several of the greatest scientific minds of Owen's time, including zoologists, botanists, and geologists, had aided the venture—by giving books to the library, collecting money, and propagandizing for the cause.

Several chronic problems faced the Owenite communists at this point. First and foremost was Owen's impetuosity itself. He was a teetotaler and did not tolerate strong drinks in the community. He baited those in the group who professed orthodox religious convictions. Class cleavages were evident also. The middle-class intellectuals were prone to resent the difficult physical labor expected of them and tended not to intermingle with those of humbler origin. A letter sent by one of the communitarians to the *New Harmony Gazette* of January 31, 1827, complained of the "slow progress of education in the community—the heavy labor, and no recompense but cold water and inferior provisions." [30] Another proclaimed, "We had bread but once a week . . . on Saturdays. I thought if I ever got out, I would kill myself eating sugar and cake." [31] Moreover, many parents began to resent Owen's educational ideas, which included separating children from parents for extended periods of time. Owen commented at one point that:

> You also know, that the chief difficulty at this time arose from the difference of opinion among the professors and teachers brought here by Mr. Maclure, relative to the education of the children, and to the consequent delay in putting any one of their systems into practice.[32]

29. Quoted in Rowland Hill Harvey, *Robert Owen* (Berkeley: University of California Press, 1949), p. 116.
30. Noyes, *op. cit.*, p. 49.
31. Harvey, *op. cit.*, p. 128.
32. *Ibid.*, p. 129.

Owen finally admitted defeat in March, 1827, and commenced a series of lectures and debates against prominent religious figures. William Maclure immediately took control of the community. As an eminent scientist and educator, he hoped to continue educational innovations within the organization. Moreover, he created a School of Industry which foresaw the pragmatism of twentieth-century "progressive education." One of his rather "heavy" educational works was entitled, "On the possibility of Improving Practical Education, by Separating the Useful from the Ornamental, and thereby Reducing the Labour and Fatigue of Instructing Youth . . ." [33] Maclure was foredoomed to failure, however, and by 1835 donated the remains of his New Harmony Library to the Academy of Natural Sciences at Philadelphia. This is sometimes referred to as the first truly public library in the United States.[34]

Noyes comments on the reasons for the failure of New Harmony:

> Owen's method of getting together the material of his community seems to us the most obvious external cause of his failure. . . . A public invitation to "the industrious and well-disposed of all nations" to come on and take possession of 30,000 acres of land and a ready-made village, leaving each one to judge as to his own industry and disposition, would insure a prompt gathering—and also a speedy scattering. . . . Judging from all our experience and observation, we should say that the two most essential requisites for the formation of successful communities, are religious principle and previous acquaintance of the members. Both of these were lacking in Owen's experiment.[35]

The Owen experiments in other parts of the country failed also; yet they had in several instances been the progenitors of such modern ideas as racial and sexual equality.[36]

FOURIERISM IN AMERICA

It was perhaps kind of fate not to permit Fourier to live to see the legion of Fourierist experiments in the United States. They seldom

33. W. H. G. Armytage, *Heavens Below: Utopian Experiments in England* (Toronto: University of Toronto Press, 1961), p. 125.
34. *Ibid.*, p. 136.
35. Noyes, *op. cit.*, p. 57.
36. *Ibid.*, pp. 71, 86.

followed the rules laid down by their ideological founder and floundered hopelessly on American soil, for the most part. John Humphrey Noyes cites thirty-one Fourierist communities in the United States, located in Massachusetts, Michigan, Iowa, Ohio, Illinois, Pennsylvania, Indiana, New York, and New Jersey.[37]

By the time of his death in October, 1837, Fourier had accomplished little in giving reality to his utopian vision. He had, however, converted two important men to his cause, M. Baudet-Dulary and Albert Brisbane. Monsieur Baudet-Dulary, in 1832, founded a newspaper with Fourierist ideas, *Le Phalanstere*, which later failed, and created the first Fourierist community in France, which was also foredoomed.

Albert Brisbane was a brilliant young man who came to Europe to study social philosophy under Hegel. After showing interest in Saint-Simonianism and other utopian ideologies, he came across Fourier's *L'Association Domestique-Agricole*, which he saw as the solution to his philosophical quandaries. After intensive study under the master himself, Brisbane returned to America intent on realizing the rather fantastic goals of Fourierism.

By 1840 Brisbane was ready to begin. He published in that same year *The Social Destiny of Man*, which distilled Fourierism into a language more suitable for attracting converts to the message than had been Fourier's rather ominous and massive works.

> It is very probable that the excitement propagated by this book turned the thoughts of Dr. Channing and the Transcendentalists toward association. . . . Other influences prepared the way. Religious liberalism and antislavery were revolutionizing the world of thought, and predisposing all lively minds to the boldest innovations. But it is evident that the positive scheme of reconstructing society came from France through Brisbane. Brook Farm, Hopedale, the Northampton Community and the Skaneateles Community struck out, each on an independent theory of social architecture; but they all obeyed a common impulse; and that impulse, so far as it came by literature, is traceable to Brisbane's importation and translation of the writings of Charles Fourier.[38]

Another major breakthrough for Brisbane was his contact with Horace Greeley, famous and influential editor of the *New York Tribune*. Brisbane purchased a twice weekly column in the paper and wrote effec-

37. Noyes, *op. cit.*
38. *Ibid.*, p. 201.

tive prose concerning "Association, or principles of a true organization of society." [39]

Brook Farm in Massachusetts was founded by intellectual and literary giants, such as Ralph Waldo Emerson and Nathaniel Hawthorne,[40] as a communal venture. Charles Dana wrote in 1841,

> At Brook Farm they [laborers] are all servants to each other; no man is master. We do freely from the love of it, with joy and thankfulness, those duties which are usually discharged by domestics. . . . Again, we are able already, not only to assign to manual labor its just rank and dignity in the scale of human occupation, but to insure it its just reward. And here also, I think, we may humbly claim that we have made some advance upon civilized society.[41]

Brisbane moved his propaganda machinery from New York to Brook Farm and so influenced the Brook Farm membership that in 1844 they changed their name from The Brook Farm Association to the Brook Farm Phalanx. In a true Fourierist manner, a joint stock company was formed. Many members took shares by paying money; others held shares by their labor. An old house on the place was enlarged and three new houses built. The Brook Farm group did not attempt to revolutionize the family structure; however, it did develop a common school and a common nursery. Ralph Waldo Emerson commented on the division of labor in the community in this way:

> In Brook Farm was this peculiarity, that there was no head. In every family is the father; in every factory, a foreman; in a shop, a master; in a boat, the skipper; but in this Farm, no authority; each was master or mistress of his or her actions; happy, hapless, anarchists.[42]

Several of the intellectuals tired of the monotonous days spent in hard labor, notwithstanding their comments concerning the dignity of hand labor. Defections in the ranks began as financial problems increased. In March, 1846, the hopes of the community were dealt an almost fatal blow as the central building on the farm was demolished

39. *Ibid.*

40. *Vide* Manning Hawthorne, "Hawthorne and Utopian Socialism," *New England Quarterly* 12 (1939): 727–29.

41. Noyes, *op. cit.*, p. 223.

42. Quoted in *Autobiography of Brook Farm*, edited by Henry W. Sams (Englewood Cliffs, N.J.: Prentice-Hall, Inc., 1958), p. 225.

by fire. George Ripley commented on the possible future of the group two weeks after the fire:

> We cannot now calculate its ultimate effect. It may prove more than we are able to bear or . . . it may serve to bind us more closely to each other, and to the holy cause to which we are devoted . . . we have every reason to rejoice in the internal condition of our Association. For the few last months, it has more nearly than ever approached the idea of a true social order.[43]

Within a year the society had dissolved and the farm was disposed of. The farm had lasted three years, which were seen as a valuable time by its membership. Yet the attempt to receive the industrial age, in an agricultural setting and with intellectuals as laborers, seems doomed from the start.

Brook Farm was only one of a profusion of Fourierist groups, nonetheless; and several others were somewhat more successful in a temporal sense. The North American Phalanx in New Jersey, for example, lasted no less than twelve years. Its location near New York City and its active support by Brisbane and Horace Greeley aided it greatly. The Phalanx was begun in 1843 with almost $8,000 in cash; by 1852 its property was estimated at ten times that value.[44] Labor was paid at six cents per hour[45] and meals were bought for various prices in the dining rooms. The North American Phalanx attempted to diversify its economic resources by brick-making, milling, and the like, and gave great support to social equality for women and other radical ideas of the day.

The religious form of the community followed the same lines as the Brook Farm Phalanx. As one visitor to the colony put it in 1847:

> There is religious worship here every Sunday, in which all those who feel disposed may join. The members of the society adhere to different religious persuasions, but do not seem to care much for the outward forms of religion.[46]

Fire was the nemesis of the North American group, just as it had been at Brook Farm. Eleven years after the founding of the society, the members of the North American Phalanx watched their mill burn down. After lengthy discussion concerning their future, some suggested that

43. *Ibid.*, p. 174.
44. Noyes, *op. cit.*, p. 461.
45. The Phalanx doctor was, of course, paid the same wage.
46. Noyes, *op. cit.*, p. 473.

they dissolve, and, surprisingly to many, a vote was taken indicating that they should dissolve. Investors in the community settled for sixty-six cents on the dollar. Thus, by the beginning of the Civil War, Fourierism, along with Owenist experiments, had passed into history. Noyes sums up the movements in this way:

> Owen's plan was based on communism; Fourier's plan was based on the joint stock principle. Both of these modes of combination exist abundantly in common society. Every family is a little example of communism, and every working partnership is an example of joint stockism. Communism creates homes, joint stockism manages business.[47]

Unfortunately, in this case, the Owenite families were basically incompatible, and the Fourierist businesses, in the end, without profit.

Redelia Brisbane, daughter of the American Fourierist Albert Brisbane, comments on the demise of her father's utopian community.

> The organization was not adapted to the natural and manifold wants of its members; the legitimate aspirations and ambitions of individuals found no satisfying field of action. . . . It was inevitable, finally, that individual members, perceiving that there existed outside of their little community a field of action more in harmony with personal requirements and ambitions, should turn their backs on the ideals of youth to mingle again with the outside world in broader and more complex spheres of action.[48]

ONEIDA—JOHN HUMPHREY NOYES

The Oneida experience is, in one sense, an epilogue to the early communalist period of American life. Based on the concepts of the highly educated and articulate John Humphrey Noyes, the Oneida colonists in 1844 adopted communism in upstate New York. Noyes had participated in antislavery agitation earlier and believed, as did the early Shakers, that Christ's second return had occurred at precisely 70 A.D. Since this momentous event had already taken place, man was free to seek absolute perfection on earth.

Noyes was moved by the statement that in heaven "they neither marry nor are given in marriage." While this led the Shakers to com-

47. *Ibid.*, p. 194.
48. Redelia Brisbane, *Albert Brisbane* (Boston: Area Publishing, 1893).

plete abstinence, Noyes moved in the other direction as he proposed his system of "male continence" and "complex marriage" which was adopted by the community in 1846.

> We are opposed [he wrote in 1847] to random procreation, which is unavoidable in the marriage system. . . . We believe that good sense and benevolence will very soon sanction and enforce the rule that women shall bear children only when they choose.[49]

Further, Noyes disparaged monogamy which "gives to sexual appetite only a scanty and monotonous allowance and so produces the natural vices of poverty, contraction of taste and stinginess or jealousy.[50]

The answer to this problem was complete freedom of intercourse combined with male continence to prevent unwanted births. Holloway states that "on the whole the system was remarkably successful." [51]

Another, rather less dionysian, innovation at Oneida was "mutual criticism" which involved the entire society as it either castigated or praised an individual for his acts. If a member of the society was seen as unduly prejudiced against the individual in question, he too was rebuked by the group.

The membership of the group was eighty-seven in 1849, but by 1878 the community had grown to over three hundred. A valuable library was established with books by Darwin, Spencer, Huxley, and others. The seventh decade of the nineteenth century brought with it some disharmony in the community. It was, typically enough, a conflict of generations. Noyes's son was an outspoken agnostic who caused a good deal of dissension within the ranks. Verbal attacks by outsiders also threatened the community at this time.

Noyes began to realize that radical change would be needed to salvage the community, and in August, 1879, he advised against further practice of complex marriage as a tactic, although he remained convinced of its essential truth. In 1881 the colony moved closer to respectability, as they abandoned communism for the joint-stock method of operation. At this point, gradual disaffection began to occur; a colony of twenty moved to California; others moved to New York City, Boston, and Niagara. In April, 1886, Noyes died, leaving only a corporation of craftsmen bearing the name Oneida.

49. Holloway, *op. cit.*, p. 184.
50. *Ibid.*, p. 185.
51. *Ibid.*, p. 186.

LESSONS FOR THE PRESENT

The causes of the failures of the early American communes are complex and not easily disposed to analysis. Many of the groups met their nemesis for ostensibly economic reasons. Most groups decried industrialization and turned to an agrarian system of values. Farming was a demanding and sometimes unrewarding profession. The capitals held by the memberships were often insufficient to promote expansion or to compensate for emergencies and disasters.

It would be incorrect to see financial problems as insurmountable to all groups, however. In fact, many were experiencing prosperity at the time of their demise. The major problem, instead, would seem to be the lack of commitment of the membership to the group's goals. Many members of social organizations have endured great poverty and deprivation as they struggled to facilitate the purposes of their organization. What, then, are the factors in the social structure making for maximum commitment?

Rosabeth Kanter gives a tentative answer to this question in her article, "Commitment and Social Organization: A Study of Commitment Mechanisms in Utopian Communities." [52] Kanter surveys ninety-one utopian communities in existence between 1780 and 1860. "Successful groups," [53] she finds were likely to exercise greater social control over their membership. How did this control operate? First, by forcing the individual to renounce the outside world by rejecting the larger society's pattern of marriage (*i.e.*, requiring some form of celibacy or even "free love"). Second, successful groups were able to develop a feeling of communion through group ritual, collective singing, or the like. Individuals belonging to these successful utopian groups were prone to great sacrifices, such as giving up pleasures of the world (alcohol, dancing, or reading). Further, Kanter finds that they made real investments, such as irreversible donations of money or property to the communal group. Strongly cohesive groups often wore common uniforms and used techniques producing "surrender" and "mortification," such as "mutual criticism."

52. Rosabeth M. Kanter. "Commitment and Social Organization: A Study of Commitment Mechanisms in Utopian Communities," *American Sociological Review* 33 (October, 1968): 499–517.

53. Successful utopian ventures are defined by Kanter as lasting twenty-five years or about one generation.

Many of these techniques, it is true, would be repellent to current communalists due to their concern for individuality. The price for stability in communal groups does seem to be the acceptance of some form of authority. As we shall see in the next chapters, however, stability is not the overriding concern of many modern communalists.

4

EROTIC UTOPIANS

"When people find out you know—that I'm living in a commune—the only thing they want to know about is the sex thing. It really bugs hell out of me."

from an interview with a twenty-year-old midwestern communalist

SEXUAL OPTIONS

In the Santa Clara mountains of California a small group of communalists practice nudism, study astrology, and "do drugs." Of the ten girls resident on the farm in 1969 six were pregnant. Polyerotic attachments were seen as normal by the communalists, and children born in this setting are regarded as belonging to all, or as "children of the commune."

This is perhaps an extreme case in the actualization of a new sexual morality by communalists, but it is a "picture in the mind" that stays with the general public. Actually, since communal groups vary as much in their attitudes as do larger societies, they may in many cases hold purely conventional attitudes toward sex and marriage. Twin Oaks, patterned on B. F. Skinner's operant conditioning principles is a case in point. This communal farm in Virginia operates on a typical monogamous family pattern. Some in the group have advocated a more complex system of sexual relations, nonetheless the strong cultural conditioning of first generation communalists makes sexual experimentation more than a little difficult.[1]

1. See "Twin Oaks," *The Mother Earth News* 1, no. 1 (January, 1970): 56–59; see also "Walden Two Lives," *The Modern Utopian* 2, no. 1 (September, 1967): 3–7.

In general, it may be safe to assume that the less a commune is structured in terms of economics, division of labor, and the rest, the less it will conform to middle-class sex norms. Since not all communes are unstructured, it would follow that not all deviate sharply from accepted modes of sexual behavior in the United States. It is, of course, impossible to measure accurately the amount of nonmarital sexual activities that occur in the modern communes. There is some evidence to support the idea that it is not as prevalent as the "silent majority" would believe.

Professors Stanley Krippner and Don Fersh visited eighteen communes in 1967, 1968, and 1969.[2] Their impressions of the amount of sexual license permitted was that only three communes granted "much" sexual license, eight granted "some," and seven allowed "little." Two of the three that allowed "much" sexual license Krippner and Fersh termed "unstructured secular communes."

REPRESSION AND SURPLUS REPRESSION

In the late 1960s, a number of observers concluded that the nuclear family in America was undergoing a monumental crisis. Due to the demands of technology the family became much more mobile and "streamlined," as uncles, cousins, and grandparents were left behind in the search for better jobs. The loss of such relatives from the household meant that individuals when they married were to become almost entirely dependent on their mates for emotional support. Some have contended that this emotional dependence has been too great for the nuclear family to bear. Certainly, any citation of divorce statistics or estimations of marital discord would lend credence to this point of view. Another indication of the inadequacy of some marital relationships is strikingly evident in the newly publicized phenomena of "mate-swapping." *Psychology Today* reported in July 1970[3] that five percent of its highly educated readers stated they had participated in mate-swapping, while a full third of the respondents admitted they might participate.[4]

2. Stanley Krippner and Don Fersh, "Mystic Communes," *The Modern Utopian* 4, no. 2 (Spring 1970): 2–10.

3. "Sex," *Psychology Today* 4, no. 2 (July, 1970): 43.

4. Mate-swapping seems to be a curious mixture of conservatism and experimentation. Many swappers give as their rationale for their activities the idea that their marriage has been strengthened. The justification of an essentially conservative institution such as marriage by such socially unapproved means is perhaps sympto-

Two of the most popular philosophical advocates of less sexual repression in current times have been Wilhelm Reich and Herbert Marcuse. Combining varying proportions of Marxism and Neo-Freudianism, Reich made two crucial hypotheses: first, "The determining factor of the mental health of a population is the condition of its natural love life,"[5] and second, "Sexual suppression is an essential instrument in the production of economic enslavement."[6] Reich, who is much read among new communalists, called for a sexual revolution. An end to premarital chastity by adolescents and an abolition of concern for marital fidelity were prerequisites for a postrevolutionary "sane society" according to the Reichian view.

Another, perhaps more erudite, but no less creative, erotic thinker has been Herbert Marcuse. Marcuse's popularity with the New Left is not open to doubt.[7] Marcuse, like Reich, came out of a tradition of German Marxist thought (in Marcuse's case with a strong Hegelian flavor) and Freudian philosophy. While Marcuse agreed with Freud that a certain amount of sexual repression was necessary for any society to function as an ongoing system, he parted company with Freud on the issue of economic repression (*i.e.* capitalism). Freud had not taken into account the tremendous force of capitalist bureaucracy in shaping the sex ethics of our society. Under the rule of capitalism ". . . body and mind are made into instruments of alienated labor; they can function as instruments only if they renounce the freedom of the libidinal subject-object which the human organism is and desires."[8]

What Marcuse is expressing here is the idea that capitalism produces within the individual a desexualization of the body in general, with the exception of the genitals. Playfulness and an oceanic sense of sexuality are foreign to capitalism, according to Marcuse. The reason for this "genital tyranny" and "surplus repression" is that the economic system channels the instinctual energy of the individual into nonsexual but

---

matic of the contradictory time in which we live. It is likely that many of the swappers are motivated by somewhat neurotic needs, since the forbidden nature of their activities, coupled with a certain degree of guilt, seems to enhance their sexual enjoyment.

5. Wilhelm Reich, *The Discovery of the Orgone: The Function of the Orgasm* (New York: Farrar, Straus and Giroux, 1942).

6. *Ibid.*

7. Leftist students in Paris recently carried banners celebrating three of their heroes—Marx, Mao, and Marcuse.

8. Herbert Marcuse, *Eros and Civilization* (New York: Beacon Press, 1955, p. 46.)

productive areas. Thus, the very nature of the totally productive-consuming society is sexual malaise.

If one takes Marcuse seriously, as a great number of communalists do, the sexual revolution must accompany or even precede radical socio-political changes.

COMMUNAL EXPERIMENTATION—GROUP MARRIAGE

How much sexual repression or control is necessary for a communal society to exist? Some communal advocates of group marriage believe they have found the key to this question. "Group marriage," writes one communalist, ". . . avoids the pitfalls of exclusive monogamy and impersonalistic promiscuity." [9] Does group marriage in fact avoid the pitfalls of monogamy and unrestrained sex? To answer this question it is important to remember that current experimenters in group marriage have grown to adulthood in a (relatively) monogamous society. They are first-generation participants in the creation of an "erotic utopia" and as such they have been imbued with many of the sexual and emotional responses programed into all of us.

"Jim," [10] a member of an east coast urban group-marriage community, describes his situation in this way.

> The house is limping along. Jane was turned off by Ted and Ruth. . . . Ted and Ruth came in as a couple deeply turned on to each other and unprepared to share closeness with others. Also, both are pretty much down on the extended family concepts and some of the other things that Jane, Betty, Susan and I were into (Ted was really caustic at one point about encounter techniques). Basically, I think they were looking for a place to live while they pursued other things. . . . The rest of us, especially Susan, Jane and I, were looking at the house as *the* thing. I think we have all learned that neither extreme is "right," but the end result is that Ted and Ruth are moving out at the end of this month. . . . Betty and Susan are getting along beautifully. We are hosting an encounter group on a regular basis starting Friday. The kids are doing fine. The food is interesting if not always delicious . . . considering how difficult this group living is, I think we are doing well. . . . I am beginning to think that a 90 to 95 percent failure

9. Wayne Gourley quoted in *The Modern Utopian* 2, no. 5 (May–June): p. 10.
10. The names of the communalists here have been changed.

rate for group marriages and a 50 to 75 percent turn-over rate (per year) for communities might not be unrealistic. If we compare this to the situation with nuclear families (considering alienated spouses and/or children, mental illness, as well as divorces as evidence of failure) it really isn't as bad as it seems and seems to me to be much more worth the risk.[11]

Jim here cites some of the conflicts of value which are common to all communalists—loyalty to an individual versus loyalty to the group, outside interests versus group-centered interests and the like. Nonetheless, Jim's conclusion is basically optimistic. Group marriage can work. Problems associated with it can be obviated. Interestingly enough, one of Jim's mates—Susan—feels differently about their situation.

Jim's thing about the house I think is fairly accurate. And yes, Betty and I are getting along well, but I'm not completely satisfied. She claims to feel closer to me than to Jim, but I obviously don't feel the closeness that she does, and I can't see that she feels close to me. I like her. She's a very positive person and very (overly) optimistic. But, every time I think of Jim even thinking of going to bed with her I get paranoid as hell. And, he has difficulty allowing me to express these feelings to him, and gets all defensive and I close up, or off or whatever it is that I do. I'm still so country that I could scream sometimes. I really haven't gotten to the point that I can analyze the feelings that I have when I feel threatened by their relationship. All I know is that I become panic-stricken and I feel that she has access to some intangible something that I feel belongs to me. I am very envious of the fact she and Jim have long discussions with each other and I feel left out, even though I know damn well I could join them and get into things too. I have caught myself playing the game of waiting for one of them to say "please come and join us, we want you with us." How stupid and childish, and I hate myself for it. And that makes me feel worse still.[12]

Thus, it seems that, for some at least, sexual repression and possessiveness do not disappear by simply formalizing a group marriage. Old values, such as sexual possessiveness, may become even more threatening in a group marriage. The key variable here seems to be that of prior possession. If two marital partners join a group-marriage situation, their

11. Quoted in *The Modern Utopian* 4, no. 2 (Spring, 1970): 13.
12. *Ibid.*, p. 14.

prior claims of exclusiveness to each other may complicate their new status. In all likelihood a number of single individuals with no prior claims to each other's emotions could avoid this dilemma.[13]

Another communal group advocating and practicing "group marriage" is the San Francisco based "Harrad West." In 1969 it contained six adults and three children, and actively sought new communal partners. "We have discovered," they assert in their formal statement of purpose, ". . . that sharing one's mate in return for sharing someone else for a time actually enhances the pair bond relationship.[14] We feel that our friendships have deepened, our capacity for warmth and understanding increased and our lives enriched as a result of this."[15]

One of the couples involved in the group was married to each other for seventeen years; another couple had lived together two years before venturing into group marriage. Even in this situation, completely laissez-faire sex relations are not the rule. One member of Harrad West found,

> . . . it works better to know ahead of time who you'll be sleeping with. We spend two or three nights a week with our pair-bond partner (original mate), one night with each other member of the opposite sex. Sex every night is not a must and spontaneity is not ruled out—especially during the day.[16]

The children of the commune "recognize first their biological parents but all the adults act as uncles and aunts, loving and taking care of them . . . we pretty much follow the Summerhillian concept of child rearing. We let the kids do pretty much as they want as long as they don't abuse their freedom."[17]

Members of Harrad West attributed their success in community-building—they had been in operation almost one year—to the fact that they all had interests external to the group. They were, in fact, able to maintain friendships and vocations outside the marital group. Moreover, they criticized hippie communes for their retreatist attitudes and unorganized ways. The adults at Harrad West rotate housework on a schedule. Moreover, the communalists display a remarkably nonradi-

13. They might face equally problematic situations, however, especially if they were not "selective" in choosing group members.

14. This, of course, is reminiscent of the rationale of many swingers and mate-swappers.

15. From a brochure published by Harrad West.

16. "Group Marriage: How It Really Works," *Sexual Freedom* 1, no. 2, p. 13.

17. *Ibid.*

cal domestic attitude in that they hire a housecleaner who does "general cleaning" once a week.

It becomes evident here that a relatively nonrepressed attitude toward sex does not necessarily involve radical or revolutionary changes in other aspects of one's life.

One of the structural problems in forming a communal marriage lies in the recruitment of individuals who share a common temperament or, at least certain core values. It may be remembered that the most successful communal ventures in nineteenth-century America attracted new members who shared common religious convictions. It is unlikely that current communal groups favoring "group marriage" would share a common messianic religious belief. Therefore, unless their solidarity is based on political activism or some ideal of service, they will be directly dependent upon each other's emotional support and mutual commitment. This makes it imperative that like-minded individuals contact each other for communal marriage ventures, since exploitative thrill-seekers may jeopardize the group's stability.

Two groups are attempting to reach like-minded individuals for potential group marriages. One, a New-York-City-based organization calling itself the "Expanded Family," contacts individuals interested in group marriage or other alternatives to the nuclear family. The Expanded Family discourages members who are interested in only sexual experimentation. "An expanded family," according to their brochure,

> can be simply a close friendship of trust and respect; it may be a convenient symbiotic arrangement involving shared outings, some mutual baby-sitting and perhaps a shared vacation. . . . It can involve friends who rent apartments in the same building, friends who set up home together—or it can be a fully-fledged commune or a group marriage.

The Expanded Family specifically rejects the idea of mate-swapping but recognizes the "centrality of sex as a major factor in human relationships."

Further, they argue that,

> when adults relate to each other on more than a most superficial level, sexual attractions and sexual tensions are bound to have an important influence. The only position the Expanded Family takes on this point is that the sexual factor must not be ignored. Sexual

attitudes and conflicts should be discussed only by those involved. In many cases adults who are involved in a strong emotional or intellectual relationship with others will decide that the relationship should also become a sexual one. Others will seek ways of sublimating this force. The decision is one to be decided personally by those involved.

In the New York area the organization holds weekly discussion groups, encounter-group sessions, and recreational outings. A most interesting facet of the organization's member contact service is the questionnaire to be filled out by prospective members. Some of the statements the couple must answer "yes" or "no" to include the following: "We would expect to be sexually free with any couple with whom we related really well." "We take care that our children don't observe us in a sexual situation." "We have participated in group therapy, sensitivity groups, etc.," "We have a professional background in the behavior sciences (give details)," and finally, "We have formed close ties with other families that involved sexual relations."

Another organization, The Alternatives Foundation, also attempts to create a communication network between communally oriented people with its communal Match Service. The Foundation has developed a rather extensive questionnaire which inquires not only as to the sexual practices and desires of the applicant but his attitudes toward drugs, politics, child rearing, and religion as well. Moreover the questionnaire has a section entitled "What turns You off?" The individual then proceeds to check those personality traits he finds intolerable in others. The effect of this is that the individual is matched with others much in the manner of computer dating. The Alternative Foundation also experimented with a "group dating" service which failed due to the fact that four times as many males as females applied for the service (leading the Foundation organizers to believe that the motives of the applicants may have been more base than utopian.)

LAISSEZ-FAIRE SEX

Neither Marcuse, Reich, or their mentor Sigmund Freud ever proposed a complete moratorium on sexual restraint. Yet today a small number of utopists have carried the idea of sexual license to its ultimate. The Sexual Freedom League, with chapters in Denver, New York City, Baltimore, Chicago, San Diego, Los Angeles, and San Francisco, is attempting to do away with all societal constraints on voluntary sex. The League founded by a New Yorker, Jefferson Poland, has both political and pragmatic aims. On the one hand it seeks to liberalize laws pertaining to sex (*i.e.* abortion, prostitution, and pornography statutes) and on the other its "nude parties" form a nucleus for sexual variety and experimentation.

One of the female members of the League describes her activities in the group in the following way:

> I have quite a few close friends, and we are likely to make love whenever the opportunity presents itself. We're almost all heads, and very hang-loose, and not the sort of people so sexually obsessed as to ever go very far out of our way to seek sexual activity, or a partner. The idea of having a classic affair seems sort of trite and almost laughable. And I really enjoy sharing my bed with another warm critter . . . and if that critter cares to initiate any sex play, I'm generally agreeable . . . Sex is fun. It's also a very enjoyable—and effective—way to tell a friend, I love you.[18]

Notwithstanding this Leaguer's advocacy of playful nonrepressive sex, the Sexual Freedom group has not grown phenomenally in the past few years. While most of its members are men, female Leaguers are steadier in their attendance at League functions.[19]

The reason for the lack of rapid growth of the movement may be due to the fact that many individuals who are relatively liberated in a sexual sense see no need to politicize their beliefs or activities. More importantly, if sexual freedom advocates are correct that sex and emotion are unrelated—it seems logical to assume that they will be unlikely to develop a sense of community around the ideology and practice of sex. Stated another way, complete abolition of sexual restraint and repression makes lasting communities or societies impossible.

While it is true that a number of hip communes do have a large measure of sexual freedom, they are generally transient in nature. As Rosabeth Kanter argues "The prospects for most of today's anarchistic communes are dim; they lack the commitment-building practices of the successful communities of the Nineteenth Century." [20]

Is it possible to conclude that the new communalists can develop a

18. "Yes If You Wish," *Sexual Freedom* 1, no. 2, p. 25.
19. See Jack Lind, "The Sexual Freedom League," *Playboy*, November, 1966.
20. Rosabeth M. Kanter, "Communes," *Psychology Today* 4, no. 2 (July 1970): 78.

less repressive sex code and still maintain a sense of social cohesion? The answer seems to be a highly qualified "yes." The family unit can be rearranged to provide potentially more satisfying configurations than the present ones. However, it must retain some of the functions of the older family units—the crucial one being support and concern for the family member.

Some of the difficulties to be faced by the new communalists in establishing experimental or group families are not to be taken lightly.[21] First it seems likely that members of a group marriage would be unable to pursue many outside careers, since they might necessitate moving around the country.[22] Possibly this problem could be obviated by the development of cooperative economic ventures by the group as a whole. Nevertheless, in many cases individuals may be forced to sacrifice career aspirations to maintain their membership in the group family unit.

Another problem of any group marriage would concern the idea of sexual possessiveness. New jealousies and conflicts are sure to develop in a group marriage.[23] It is more likely, as I have suggested, that individuals coming into the group marriage with no prior commitment to sexual exclusiveness would be less threatened by the group marriage structure. The stress and strains of adapting from the monogamous marital position to one of less sexual and emotional exclusiveness are without doubt difficult barriers to surmount.

Other real problems such as methods of relating to children, group decisions, selectivity of members (perhaps new methods of divorce as well), and the like, are all formidable at present. Nonetheless as long as technological "progress" changes the character of social relations in this country, one can expect experimentation in family life to continue at an even more rapid pace.

21. See Sharon and Thomas W. McKern, "Will Group Marriage Catch On?" *Sexology*, June 1970, p. 31–34.

22. It is difficult enough for the monogamous family to move constantly. It is nearly impossible to conceive of group marriages moving about to further its members' career.

23. For a social-psychological account of the Mormon plural marriage system and its problems, see Kimball Young, *Isn't One Wife Enough?* (New York: Holt,Rhinehart and Winston, 1954).

5

HIP COMMUNES—BACK TO NATURE

"Take all the apples you want. This isn't our land. It's God's."

☺ *a Morningstar communalist*

WEST COAST COMMUNES

Tolstoy Farm. Hip communes are notoriously unstable. Two years is an exceptional length of time for their existence. Yet, near Davenport, Washington, an anarchist commune has existed and, in many ways, thrived to the present. In a letter to the author, one of its inhabitants describes Tolstoy Farm:

> [It] is not so much a community as it is a place that is not owned by anyone—so anybody can live here if they find room and the means. There are about 200 acres in a deep canyon with about 40 people living on it. There is quite a bit of cooperation among people, considering that it is all voluntary. There are some co-op gardens, livestock, vehicles, and a co-op school. There are also private gardens, vehicles, livestock, etc. Each family has its own house, money, and whatever. We run our lives without rules and the school as well. Students can study whatever they want, and teachers teach whatever they want. There is a craft shop in the schoolhouse, and we have a crafts marketing co-op.
>
> We have get togethers (voluntary) to discuss feelings, ideas, be creative, and to sing or dance, and some potluck dinners. People can live here quite cheaply by growing food, building shelter with free materials, cutting firewood, and sharing. It is difficult to earn money while living here. This kind of life requires many changes,

especially toward self-reliance and accepting responsibility for oneself (Thoreau, Emerson).

Our school is very important in keeping us growing and involved in the world culture.

It is very beautiful and peaceful here, but it is difficult for people to learn to live harmoniously. There is room for those who are willing and able to use it. . . ."

Tolstoy Farm is a remarkable success story because of its survival and stability. Although its inhabitants live in a community "without rules," one can be sure that its more stable members have carried "internal" rules and concern for their fellow members. Unfortunately, the Tolstoy Farm experience may be atypical.

Morningstar. Sonoma County, north of San Francisco, approximates the Garden of Eden in at least two ways—climate and natural resources. A few miles from the town of Sebastapol lies the Morningstar commune. In the spring of 1966, Lou Gottlieb proclaimed his thirty acre Morningstar property "open land." Since Haight-Ashbury was on the decline for its hip populace at the time, several hundred of its inhabitants drifted into Morningstar to escape the frenetic urban scene and to "get into nature." An orchard near the top of the hills at Morningstar is lush with all manner of fruit trees. Most of the year, apples and pears lie rotting in abundance beneath the trees. The climate, too, is little short of idyllic. It is true that it rains a good deal during the winter; nevertheless, severe weather is no problem.

The dirt, clay, and rock road leading to the top of the Morningstar land is impassable for any vehicle without four-wheel drive, and with good reason. Morningstar inhabitants have suffered an unusual number of "hassles" from the Sonoma County authorities. The commune has no running water and no sanitary facilities.[1] Moreover, since Gottlieb specifically refrained from putting down any rules or regulations, the use of soft drugs, such as "hash" and "pot," has been frequent at the commune. This, with the lack of sanitary facilities, and the unreasoning fear of the straight community, has led to a number of raids, "busts," and legal attacks on Gottlieb's land—or rather "God's" land. For in 1968, Gottlieb deeded the land to "God," hoping to avoid legal

1. A friend of the author once made the mistake of asking a Morningstar communalist about the location of sanitary facilities. "The whole thirty acres is a toilet," she replied. And, as an afterthought, "You should cover it up though." She then proceeded to tell us that the two basic advantages of this method were that one "replenished Mother Earth" and "avoided hemorrhoids."

harassment.[2] By this time, however, most inhabitants of the commune had moved on to other communes in the area, such as Wheeler's Ranch. When I visited Morningstar in the late summer of 1970, only thirty residents remained scattered over the same number of acres. Tepees, shacks, tree houses, and tents provided shelter for some of the communalists, while others simply slept on mattresses and in sleeping bags.

A large, bare-chested, Mohawk-haired, young man told me that Gottlieb had confounded the local authorities to the extent that they did not know how to proceed in attacking the commune. "So for the last few months none of them have been up here. Lou knows how to handle them."

At the time of my visit, Lou was washing Vishnu's diapers. Vishnu, about six months old, was nude, as was his mother, Nooma. Morningstar was Nooma's fourth commune. "I just bug hell out of people when they get to know me." Nooma had dropped out from a Jewish home in the East. Her sister was at Wheeler Ranch, as far as she knew. "These are our new psychos," she mused and pointed to a sixteen-year-old boy and girl who looked not at all "psycho" but more than a little shy.

Down the path, nearer the orchard, a family with young parents, in their early twenties, drank beer, ate apples and pears, and, occasionally, passed around a psychedelic pipe. Their baby, fat, content, and seemingly healthy, slept while her older brother, about five, ran about showing the other children his personal hiding places. "I'm from Los Angeles," commented the young man,

> and just about all the guys I hung around with are doing hard drugs now. They'll never get their head together living down there. I couldn't take it either. But I can tell you one thing, the straight world out there is too much—nobody's together, and when it blows—that's all—but we'll be ready up here. We're learning to survive. Survive, hell! We're learning how to live. Right now I'm learning herbs and plants and stuff.

Morningstar is not a commune in an absolute sense since everyone "does his own thing" where and when he pleases. Yet there is a great deal of sharing and mutual help as the occasion arises.[3] There is a

2. In what must be one of the strangest lawsuits in the history of the judicial system, a woman, who had had property damaged in what was legally termed "an act of God," proceeded to sue God for possession of Morningstar.

3. Interestingly, a fellow sociologist, Lewis Yablonsky, visited Morningstar in

concern for the development of an ideology or world view at Morningstar and some of the other communes in the Sonoma County area. Bill Wheeler of the Wheeler Ranch and Lou Gottlieb are now attempting to create a new religious view which will give a sense of unity and purpose to the communal movement. The "Morningstar Faith" or the "Open Land Movement" has as its purpose "the living of a primitive life in harmony with revealed Divine Law." Revealed Divine Law in this case seems to be an amalgam of Buddist, Yoga, Druid, and panthiest beliefs, in short, a shamanistic view of earth and existence. The faiths, "Four Missions of Planetary Purity," "Open Earth," "Open Air," "Open Fire," and "Open Water," sound as if they had been created by a mystical ecologist.

> 1. Open Earth: the opening of lands as sanctuaries for the one, naked, nameless, and homeless. The purification of the land by replenishment of the soil, invisible land use, and voluntary primivism. . . . 2. Open Air: the opening of all communications channels . . . the purification of the air by planting groves and gardens. 3. Open Fire: opening of all energies to all. No exclusive ownership of gas, electricity, coal, wood, money. The purification of life forces in the fire of worship. The open hearth. . . . 4. Open Water: Open seas, oceans, rivers, streams, and hot springs. . . . The purification of water. Stop pollution.[4]

The Morningstar faith has its set of prohibitions, albeit vague, as do older religions. To be avoided are actions which promote the following:

> 1. Hurting for the sake of hurting; 2. Exploitation of the elements; 3. Un-invisible (plush) living; 4. Conditioning anything or anyone to your beliefs; 5. Rigid planning ahead; 6. Self-poisoning—eating poisoned food, breathing poisoned air . . . drinking poisoned water; 7. Repression of urges, feelings, instinctual drives. . . .[5]

1968. Yablonsky's perceptions of the commune were nearly antithetical to mine. He perceived fear, potential violence, inattention to children, and drug abuse as rampant at Morningstar. A time lapse of two years between Yablonsky's visit and mine may partially explain the differences in our perceptions. Some of the communalists' negative attitudes toward Yablonsky may have been precipitated by the fact that he carried a briefcase with tape recorder. See Chapter 10, "The Morningstar Bummer" in Yablonsky's *The Hippie Trip* (New York: Pegasus, 1968).

4. From a mimeographed tract.

5. *Ibid.*

"Voluntary primitivism could only evolve within an economy of abundance, such as the United States today," argue the founders of the movement. "It proposes a synthesis of the technologically sophisticated life style with a voluntary return to the ancient tested ways—living close to God's Nature and in harmony with the elements.[6]

The turnover of the population of Morningstar is tremendous, as is its effect on other communal groups. For example, one year after the establishment of Morningstar, about thirty of its inhabitants moved to Taos County, New Mexico, where they founded Morningstar East. Morningstar East, like its antecedent, is "free land" open to all. Also like its parent organization, it has no formal leadership or structures for decision making. Its highly fluid membership does attempt farming on the less fertile New Mexico soil, however. At the present time, some 45 to 50 people populate Morningstar East. Most of them are original members of the Sonoma County experiment.

Wheeler Ranch. I drove not more than a dozen miles from Morningstar to Wheeler Ranch. Founded a year later than Gottlieb's Morningstar, Wheeler Ranch is nearly 350 acres of rolling hills, forests, canyons, and communal living. A sign near the entrance reads, somewhat contradictorily, "No Trespassing" and "Please Close Gate." A drive of about three miles farther found a road, rutted and nearly impassable, which led to another fence. A sign warned "Absolutely No Dogs!" and, somewhat predictably, a couple farther down the trail appeared with a large, nondescript mongrel. Cars fortunate enough to have survived the road into Wheeler parked along the trail. License plates from New York, Tennessee, Oregon, Illinois, and Ohio testified to the underground fame of Wheeler's communal establishment. Whole families with backpacks, single "heads," and "freeks" moved slowly down the canyon trail to the main communal settlement.

In 1967, a year after the opening of Morningstar, Bill Wheeler declared his land "free" [7] for all who wished to drop out of the larger society. Wheeler, like Gottlieb, became a target for the local "powers that be." He did make some concessions to the political establishment, however, as he put in sanitary facilities of the nonflush variety. Notwithstanding these efforts, Wheeler is the constant focus of legal at-

6. *Ibid.*

7. A Morningstar communalist complained to me that Wheeler had reserved one building on the ranch for his private art studio. "If land is free, you can't save one part and say it's free all but what I want."

tacks, law suits, and investigations by narcotics agents and juvenile authorities.

The two hundred inhabitants of Wheeler's Free Ranch fall into two basic categories: serious communalists who believe they have found a supportive "home" outside the straight world, and those who drop out for a summer, to return to school or jobs with the beginning of the winter rains. One inhabitant told me that some individuals in the ranch belonged to "cliques" and, while they were not unfriendly to others, associated basically with people they knew before coming to the ranch. "The only time we really all get together is for this religious thing." [the Open Land Church]

Those who are serious about committing themselves to the communal scene are, of course, concerned about economic realities. Several well-tended gardens are scattered about the commune, and any day finds dozens of communalists hoeing, weeding, and generally caring for the food crops. A sign near a hand pump at the ranch tells visitors and communalists to conserve water since it is needed for the crops. Nearby, a fence post holds a sign giving instructions for the conservation and use of human fertilizer.

According to Mary Moore, a social caseworker in the area, a number of Wheeler people do receive either unemployment insurance or state welfare.[8] Most of the communalists who were serious about making Wheeler their permanent home saw less need for aid from the outside as their knowledge of the land and their "natural farming" abilities increased.

Most of the dwellings on the ranch, with the exception of Bill Wheeler's studio, give the impression of impermanence—pup tents, tents made from parachutes, camping vehicles, and the like. The reason for much of the seeming impermanence of the domiciles is the moderate climate and the difficulty of packing needed equipment up the hills and mountains.

Dress is more than a little informal at Wheeler's. Nudity is not *de rigueur* at the ranch, although about one fourth to one third of the inhabitants are nude at any one time, especially those working in gardens. Nudity is not so much associated with sexuality at Wheeler as it is "getting back to Nature." The young people wear their nudity with absolute naturalness, and, typically, all but the most repressed visitor reacts the same way. Although sexual activities at the commune are not

8. This was not the case with Morningstar.

always sanctioned by marriage, they are not, as some outsiders believe, completely promiscuous. Nevertheless, the predominance of nudity at the commune contributes to the stereotype many local people have of the commune as a haven of debauchery.

In fact the relations with the ranchers and farmers of the area have been one of the sources of greatest difficulty for the communalists. This was brought to my attention in a very forceful way when I was returning with friends to my car after a visit to the ranch. I found air let out of two of my tires. I had inadvertently parked on the land of a farmer, whose dislike for the hippies was legendary. One of the hip communalists consoled me by mentioning how lucky I was to escape with only two flat tires. "He shot the last car that parked there full of 30-06 shells, and the one before that he took off the emergency brake and pushed down a ravine." [9] A young Black man who lived in the ranch from time to time mused that "the next time that old man's cows come on our place, we're gonna have steaks for a week."

NEW MEXICO–COLORADO COMMUNALISTS

While Wheeler Ranch was beginning in 1967, southern California was producing a zany, somewhat bazaar, communal venture—the Hog Farm. A number of societal dropouts organized the commune on mountainous land a few miles north of Los Angeles. A local farmer had loaned his land to the group with the stipulation that they care for his pigs. The counterculturists were quick to see in the swine a symbol of the rejection of bourgeois society, hence the idea of "hogness" and "hog consciousness." The communal family, which numbered about thirty, emulated their chosen symbol by making "garbage runs" on the nearby villages—that is, looking for edibles in the refuse thrown out by grocery stores and restaurants. The Hog Farmers' chosen symbol and behavior ignited more than the usual ire against longhairs, and, less than a year after the communal family had begun, its members were taking to the road in wildly-colored, psychedelic buses.[10] Many drifted across country and lost touch with the group. Some, however, drifted into northern Mexico.

Taos County soon became even more of a communal haven than

9. Several of the communalists made a number of trips over the horrible road for a tire pump—an act of kindness the author will not soon forget.

10. Hog Farmers drove their bus, "Road Hog," to the Woodstock Rock Festival, where they were hired as part of the "Security Please Force."

Sonoma County, California. Unlike northern California, northern New Mexico is arid, nonbountiful terrain. Even so, communes of all varieties began to spring up, including Lorien, with about thirty members; New Buffalo, about the same size; Lama Foundation, with twenty-five members; Morningstar, with over fifty residents; and The Reality Construction Company, with twenty-five members. One realtor in Taos estimated that hip communalists had spent between $300,000 and $500,000 for land in the area.[11]

New Buffalo is perhaps typical of the communal scene in New Mexico. Its members cultivate vegetables as well as grains, such as wheat and corn. Unlike some communalists at Wheeler, no one at New Buffalo subsists on welfare. Farm work is difficult for the communalists, especially when it is done by young men and women, who must learn even the most basic farming techniques. Of course, "Buffalos" distain the use of pesticides and "unnatural" fertilizers. "Organic" or "natural" foods are either raised or purchased in the bulk. Perhaps because of the harsh demands of the New Mexico soil and climate, the Buffalos seem to be more disciplined than their California counterparts. Moreover, New Buffalo communitarians have begun to limit the size of their communal family—an action Gottlieb and Wheeler have thus far avoided in the California communes. The sheer size of the influx of hip communalists has caused problems for established groups such as New Buffalo. In April of 1970, *The Albuquerque Journal*[12] estimated more than 1,700 "hippies" in the Taos area. Far more than this were expected in the summer of 1970, as schools and colleges recessed.

A member of the Reality Construction Company, a commune about five miles from New Buffalo, advised hip visitors to stay in their own part of the country to fulfill their Utopian dreams. "Don't come to New Mexico, and if you are already here with nothing to do—leave!"

The situation in New Mexico was exacerbated by the fact that in 1969 two young women were raped by men they described as "hippies." The outrage of the straight populace was predictable and immediate.[13]

11. See William Hedgepeth and Dennis Stock, *The Alternative: Communal Life in New America* (New York: Macmillan Company, 1970), p. 78.

12. April 9, 1970.

13. On April 3, 1970, a Volkswagen van belonging to a hip family was dynamited near Penasco, New Mexico. On April 8, four drunken men beat up several hippies on the street. The drunks were fined only 5 dollars each. The same night a hip resident had his home stoned, and was beaten up when he came out to protest. A few days later, in Taos, a photographer with long hair was beaten and kicked

Reports flourished in local papers about Taos housewives carrying guns to do marketing. The Charles Manson murder case added fuel to the flames, since Manson's long hair and unorthodox behavior were labeled "hippie" by the press.

When one considers the social and political climate of Taos County, it becomes a miracle indeed that the "nature freaks" in the area are surviving and, in some cases, thriving.

Only a few miles from the New Mexican border, near Trindad, Colorado, lies Drop City, another "early" communal organization. Originally started by dropouts from the University of Colorado, the group was almost strangled by "success." After the group had been "lionized" by a national magazine, it experienced an influx of hip and not-so-hip visitors, which ruined the privacy and ability to "do your own thing" in the "City." Most of the members of this anarchist commune are artists or musicians. The most spectacular art forms in the commune are the geodesic domed shelters. With axes and hacksaws, the "Droppers" have cut the tops from junked cars and nailed them to wooden joints which approximate architect-futurist Buckminster Fuller's famous design.

A Drop City communalist wrote me the following obscure lines which indicate the mood, if not the explicit nature, of the commune. "There is almost nothing interesting here but the goats—three new ones, and all that appears to be here quite a maze at least—complex geodomic cartop life. . . . If you come [to the commune] bring your own headstone with you."

Further, "Jack" attacked in an equally obscure way the obessions of middle-class America and concluded with the following: "So that seems to be our [the larger societies'] main illusion—money, friend or fiend, with or without, our policy is absolute nothingness—rule by the woodchucks."

"Rule by the woodchucks" may well be the ultimate in anarchy. A cynic might be led to comment that this kind of rule would explain the fact that only one of the original founders remains in the commune at this time. Be that as it may, the above quotation does show a desire for quiescence that is common to many hip communes.

---

after he was accused of being a hippie by six Chicanos. Local police have been decidedly nonchalant in reacting to violence against longhairs.

See Jon Steward, "Hunting Hippies in New Mexico," *Scanlans Monthly* (September, 1970): p. 23–33.)

CANADIAN EXPERIMENTS

For a goodly number of young Americans, present-day Canada holds the same image of escape as did California of the last century. Land is cheap in remote parts of Ontario, and the prairie provinces, and communal advocates have begun to take advantage of this fact. Contacts between Canadian communalists in British Columbia, Ontario, and points in between, are maintained by magazines such as *Alternate Society*[14] and a hip mail-order catalogue[15] published by "The General Store" in British Columbia. Most Canadian communalists of the hip variety share the same life style and goals as their American counterparts. They wish to get back to the land, to develop a sense of self-sufficiency, and to reject the bourgeois notion of continual, ever-increasing, consumption.

Morning Glory farm, in eastern Ontario, is one hundred acres of fir trees, rich soil, and small gardens. The communalist who bought and developed the land paid less than $4,500 for it. Four other communal farms nearby repeat the pattern. Organic gardening (nude in the summer) is the order of the day. In this area of Ontario, where land is plentiful and cheap, the communalists often exchange labor with each other in a manner reminiscent of the old time "house raising" on the American frontier.[16]

The villages in the area are not hostile to the hip communalists at this time. In fact, local businessmen have come to look to the hip population for a good deal of trade. No incidents have been reported between the police and the long hairs at this point either.[17]

This is not to say that all the communal experiments in the area have succeeded; to the contrary, another communal venture nearby was a total failure. Opposed to the Vietnam War and prone to social experimentation, a Canadian woman turned her 220 acre tract of land, complete with buildings, over to American expatriots who had fled the country to avoid military service. Since "Alice," the owner of the farm, laid down no hard and fast rules for her free tenants, they proceeded, with her away at work, to make a general mess of her home, farm

14. Published in St. Catharines, Ontario.
15. *Whole Life Catalogue* (Surrey British Columbia: The General Store, 1970.)
16. See Rod MacDougall, "Morning Glory Farm" *The Mother Earth News*, January, 1970, p. 45.
17. See "Madawaska Valley," *Alternate Society*, July, 1970, pp. 19–21.

equipment, and land. Open land and anarchy in this case led to community-destruction rather than community-building.

Western British Columbia's climate, like that of northern California, is semi-Mediterranean. Also, like northern California, it has breathtaking scenery and an increasing number of hip communalists. A summer's drive through the area reveals legions of longhairs and freaks "on a nature trip" hitchhiking and strolling along the highways. Some decide to stay and build communities. This is not easy in the area. The Social Credit party, a rightist organization, governs British Columbia, and its leadership has no love for hippies of the wandering or communal variety. The Victory commune in western British Columbia has been attacked by the provincial government for its supposedly illegal activities. Americans living there were accused of entering the country illegally, and, generally, life was made miserable for the unwanted counterculturists. The Victory commune soon went underground. "Our goals and function have been altered," wrote a member of the group.

> Beginning this summer, we still serve as an underground railway terminus for refugees from busted communes, American war resisters, etc. We refuse to harbor fugitives from the law, for our own preservation. In exchange for this underground activity, we expect to receive financial aid from underground refugee organizations, hard work from the refugees, and total anonymity. We are in a position where we may be able to offer certain vegetables and some livestock in exchange for items which cannot be grown or obtained here. . . .[18]

The movement for intentional communities in Canada is a mixture of success and loss, repression and freedom. Above all, the communalists in the great North Country seem to realize that their core problems are almost exactly the same as those of their compatriots south of the border. Organizing, experimenting, and, as often as not, failing in their endeavors, Canadian communalists are, nonetheless, tenacious. Most observers feel that the period of communal experimentation in Canada has just begun.

OTHER HIP COMMUNES

The great heartland of the United States, the Midwest, seldom begins social trends. This is the case with communal development as well.

18. "Letter from B.C." *Alternate Society*, July, 1970, p. 6.

This should not be taken to mean that the movement is not making an impact on the area; it is. College campuses in the Midwest are only slightly more conservative than those on the east and west coasts. Long hair, braless females, and highly individual dress all indicate that the bourgeois mentality is out of fashion with many of the young. Hip communes of varying structure are arising in this area as a sort of fraternity system for nonstraight students. They provide a kind of support for the nonconformist[19] student populace.

Cooperative housing is one of the mildest expressions of the communal movement on the campus scene. A twenty-year-old college student living in a Minneapolis commune told me that cooperative housing had its frustrations but was undoubtedly worthwhile. The other residents of the communal experiment did "get on her nerves" occasionally, but many of the problems of the group were solved, or at least accommodated, by the sensitivity sessions held nearly every night: "We don't have a leader. No one tries to guide the group within a certain framework. The greatest thing about our group is the absence of experts in it."

Another Minnesota commune of a far more radical nature, Freefolk, recently met its demise after a few months of struggle with primitive living conditions and Minnesota winters. Patsy Richardson attributed the breakup of the little group to a lack of privacy (the group could afford to heat only one shack during the winter, which produced what locals call "cabin fever"[20]) and to the absence of personal, creative challenges to communal members. The group had been able to maintain an open and friendly relationship with their "straight" neighbors and was not in economic trouble. It was, however, unable to develop decision-making processes that all could accept, and because of that, its members began to feel trapped and restricted by each other. The result of this was, of course, the disintegration of the community.

A reporter from the *Saint Paul Press* visited a rural hip commune near

19. It is sometimes maintained that hip young people are as conformist as any other segment of society. This is both a banal statement and a misunderstanding of the concept of conformity. If the advocate of this statement means that the behavior of the dissenters is learned, it is self-evident and a tautology. Conformity must be related to the dominant norms of a given society. Young people are stretching, changing, and experimenting with those norms; hence the term "nonconformist" is appropriate for the dissidents.

20. Most of the group's time in the winter was spent in a 10′-by-20′ community room.

the Minneapolis area. He describes the group's controversy over non-scientific farming in this way.

> The time has arrived to plant the garden . . . but there is dissension. "We should wait until Sunday," Sue declares. "We're in a fire sign now." A general discussion ensues about the agricultural applications of astrology. Most present agree that perhaps the zodiacal signs are less important than the personal vibrations of the people who do the planting. "Bad vibes, shriveled lettuce." [21]

There are yet other functioning hip communes in the Minneapolis area, such as Head and its crafts-arts branch, the Georgeville Trading Post.

Other areas of the Midwest, such as Illinois, Iowa, and Ohio, are also developing communal experiments rapidly. Usually located near a university, the communalists often get together via a local underground newspaper. In fact, a number of communalists publish their own avant-garde antiestablishment news.

Another rich source of potential communalists is the campus ministry of many protestant denominations. To revive flagging student interest in organized religion, some religious groups are beginning to experiment with a comparatively mild form of communal living.

On the east coast of the United States, the switchboard for communal communication is the Heathcote School of Living in Freeland, Maryland. The Heathcote School offers "institutes" in "Grass Roots Survival," "Changing to a Decentralized Society," "Sexuality, Roles, and Family in Community," and other communal topics. Moreover, the Heathcote School publishes *The Green Revolution,* which disseminates news of communalists, along with practical advice on chicken-raising, weaving, ecological problems, the cultivation of herbs, and the building of primitive shelters.

The members of the Heathcote community have cultivated good relations with the townsfolk. In fact, the county did not use its legal powers to shut down the organization even when it failed to meet zoning and building-code requirements. Members of the community promised to work immediately to rectify the situation.

"Lots of time and energy goes into just keeping the place going," writes Larry Lack, a member of the Heathcote staff.

21. "Commune: Living Naturally or Hip Suburbia?" *Saint Paul Pioneer Press,* July 12, 1970, p. 6.

> The visitors, 1,500, roughly, last year, whom we consider very important to our purposes here, nevertheless keep the place in constant chaos. Meals, shopping, wood gathering, and care of the animals (25 chickens and 2 pigs) are all time-consuming. Much of the time we're plagued with transportation difficulties.[22]

Despite the harsh winters and the lack of readily available land, there are perhaps a dozen rural hip communes in Massachusetts. The same pattern repeats itself in nearly all of the eastern and New England states.

Even southern campuses such as Florida State, Louisiana State, and the University of Texas were developing communal forms as the seventh decade of the twentieth century began. The South's aversion to social change has been overcome by significant numbers of its own sons and daughters, as soft drugs, psychedelic rock, antiwar rallies, and communal living became significant topics of discussion.

THE HIP ALTERNATIVE

Hip communalists embody a series of nearly endless paradoxes. While some study Department of Agriculture bulletins on scientific farming, others consult the stars for astrological counsel. In some cases, the same individual may accept the validity of both sources. Perhaps the reason for this seeming contradiction is that many communalists lack a coherent ideology. Paradoxical attitudes toward rationality and mysticism are in part due to a general characteristic common to many in the hip culture—playfulness. A great many nonconformists in our society are gifted with creativity. Innovation, love of paradox, and dialectical thinking are common to the creative dissenters. Some communalists are proud of the paradoxical nature of their ideas. In part this is due to their desire to "put on" the straight world. Whatever the cause, the creative urges of the communalists continue to be a symbol of their self-proclaimed freedom and their antagonism to the idea that one can buy an identity by driving the proper car, using the proper toothpaste, and wearing the latest Sears fashion.

Another, rather more serious, dilemma of the hip communalist revolves around the issue of "open land." The essence of the Morningstar idea was that land should be open and free to all as a permanent "drop-in"

22. "Wanted at Heathcote: A Building Fund and a Builder," *The Green Revolution* 8, nos. 3–4 (April, 1970): 2.

center. Yet the trend in many communal experiments is to close the land to new members and sometimes to visitors as well. It seems that the open-land concept would work well only with lush, fertile, climatically-mild land. Moreover, hostility from neighbors tends to put nearly unbearable pressure on this system. What is even more frustrating to permanent members of open-land communes is that they seldom get to know their communal neighbors. Continuity is necessary for the establishment of a true community. The coming and going of a large portion of the group necessitates a constant reevaluation of the commune's philosophy. Sensitivity sessions and encounter groups formed to produce knowledge and trust of others in the group are nearly worthless if new members keep appearing at meetings. Little progress toward group cohesion can ever be made in this way. Yet many open communes are reluctant to close their gates to new dropouts. "They were open to me when I came here. I know we're overcrowded, but I still can't see turning other people out," complained one communalist.

It may well be that open-land communes will never be able to establish permanent communities. This does not necessarily mean they are valueless to the participants. One important function of the open-land commune is that it places almost no demands on its young inhabitants. Adolescents in American society are faced with more crucial decisions about marriage, the draft, work, and their personal identity than any preceding generation. In many instances they are given such conflicting information about themselves and the world that a crisis of beliefs is almost inescapable. Erik Erikson calls attention to this "identity crisis"[23] of the young, and proposes that adolescents be given a moratorium on decision-making, a time to be free of responsibility and pressure. The free-land communes provide exactly that. The young people who drop in to Wheeler's Ranch, Morningstar, or Drop City are looking for something beyond what they have. If you were to ask them where they are going, they would assure you, they do not know. They are on the move in a pressure cooker society that demands answers of them, answers they cannot give. A few inhabitants of the open-land communes drift back into straight society. Most, however, seem to go on searching. For some, the moratorium on decision-making may last a lifetime.

It was said by some that the Woodstock Rock Festival in 1969, with its hard rock, soft drugs, and hundreds of thousands of hip young

23. E. H. Erikson, *Childhood and Society* (New York: Norton, 1945).

people, created "instant community." Certainly, the fact that a half million human beings could survive, yes, even cooperate with each other under conditions that included lack of food, water, shelter, and sanitary facilities, is a minor miracle. A community, nonetheless, is more than a sense of euphoria and emotional support. Communities must provide psychological support, but beyond this they must provide for group survival. Only a few months after Woodstock, a rock festival near Altamont, California resulted in tragedy. Someone with a macabre sense of humor had hired the Hell's Angels to keep order, and, as the Rolling Stones played on, the Angels beat senseless several spectators. As the huge crowd looked on, a black man was murdered. This shattered whatever illusion of community had existed at Altamont.

Many hip communes will never becomes communities in a true sense, notwithstanding their openness and propensity toward sharing. Others, such as those in New Mexico fighting for their survival, may well be on the road to community. Survival necessitates cooperation, order, and, sometimes, sacrifice. This is precisely the meaning of community.

Some would fault hip communalists for not living up to their ideals (for example, the concern for ecology shown by hip communalists living near their own litter and refuse). But such criticism can be made against any society, new or old. The hip communalists are experimenting, and, for the most part, learning from their experiments. In this way are in the best of the American tradition.

6

EASTERN MYSTICS AND CHRISTIAN RADICALS

"And the multitude of them that believed were of one heart and one soul; neither said any of them that the things which he possessed was his own; but they had all things common."

☺ *Acts 5:32*

"Whoever desires little is easily satisfied."

☺ *Lao Tzu*

LAMA FOUNDATION

One of the most common reactions to so-called hippie communes is that they are chaotic. A visit to the Lama Foundation in the Sangre de Cristo mountains of New Mexico would completely dispel that illusion. The several dozen communalists who live and work here are freaky-looking. Beyond that, their resemblances to the "do-your-own-thing" stereotype of communalists ends abruptly. Lama members are committed to a quest for self-knowledge. Self-knowledge, they believe, can best be gained without drugs or psychoanalysis. Like many others of their generation, the young Lama communalists look to eastern mysticism as a partial answer to their questions. Eastern mysticism is not a tangent for them; it is a rigorous discipline. The Lama Foundation demands work of its members in three areas—the physical, the emotional, and the mental. It is not a colony of dropouts. In fact, Lama members usually arise at 5:30 a.m. for mediation; physical exercise and chanting follow; and two hours later comes a vegetarian breakfast. Each adult works a more-than-eight-hour day, gardening, building dwellings, or fixing tools.

The foundation was started with a $20,000 grant to Steve Durkee in 1967. Durkee bought 115 acres in a 9,000 foot mountain range of northern New Mexico. The foundation has grown since then, and its members estimate that more than 300 people have lived there. By 1970, the community had a waiting list for membership. A very thorough biography of a potential member is required for consideration.

> Between the fall and winter solstice the first experimental ashram[1] was held at Lama. There were 21 communicants (13 monks, 7 karma yogis, 1 teacher); 11 buildings had been winterized; and we had use of a ranch in the valley for children. These buildings made it possible for 6 of the 20 participants to be in solitude in individual hermitages at any one time. Each monk spent 18 days in solitude and each karma yogi 9 days. In addition to the hermitage, communicants remained in silence, took daily vegetarian prasad together, did several hours of karma yoga, shared readings . . . as well as formal meditation sittings, chanted Kirtan and Mantra. There was one fast day a week (most participants fasted for periods up to nine days while in hermitage). To attempt to assess such an experience by an immediately obvious external criteria is obviously naive. The individual differences in the spiritual needs of the communicants were diverse, and in some ways the looseness of the schedule permitted each one to work in the manner deemed most suitable. Without exception the participants reported that subjectively they felt as if they had profited by the experience.[2]

The Communalists attempt to solve their problems on the basis of consensus. Majority rule is not enough. If, for example, one member dissents in a decision, no action is taken until he consents.

The physical needs of the communalists are met by doing body movement exercises such as *tai chi chuan* or *hatha* yoga. More than this, the Lama seekers do much physical labor (even on the day, once a week, in which they fast).

> Building at Lama is our medium for teaching and learning many things on many levels: trust, care, practicality, tool skills, and technics, structure realities, patience, craftsmanship, and the art of hard work and the real fact of its usability; live in it, work in it, meditate in it, eat in it, walk on it, sleep in it. Shelter! In 1969 we

1. Hermitage or retreat.
2. From a pamphlet published by the Foundation.

built a barn, 3 new domes, finished 7 "A" frames for year round use. The meeting house was completed, as well as the library and dining hall and solar heated greenhouse.[3]

Trying to scratch a living from the topsoilless mountains is no easy task. The growing season in northern New Mexico is only about 100 days. Winter temperatures fall to 30 below zero. Yet the Lama workers grow their own wheat, make their own bread, and are working on the construction of a solar-heated barn.

One of the most interesting innovations at the foundation is what the communalists call a "growhole." The growhole is dug more than five-feet into the south-southwest slope of the mountain. Two .016-inch vinyl skins cover the hole, letting the sun's rays enter the hole and keeping the severe cold out. Plants are placed in the hole near green manure, which produces a heat reaction by its decomposition. In February of 1970, green sprouts were growing in the growhole.

Besides the building of domes, "zomes" (modified geodesic domes), and all manner of dwellings, the Lama workers are attempting landscaping. A twenty-foot well is being dug; a windmill is being built. The Lama communalists believe they can construct a moneyless subsistence-economy. Subsistence is enough, according to their philosophy, and more concern for surplus things would produce the ugly spectre of materialism. The root causes of man's alienation and rootlessness are not material, according to the Lamans.

"The Lama Foundation as a spiritual communal entity is serving as an important model in the West," explains one of the group. "I think those of us that have been affiliated with Lama have a deep commitment to bring to it the highest essence of our own *beings* with which we are in tune." [4]

AN URBAN CHRISTIAN COMMUNE— REBA PLACE FELLOWSHIP

Although Christianity produces some very conservative men, it also produces radicals. In the fifteenth century, for example, Anabaptist peasants rejected the official Lutheran and Catholic churches and swept through the low country of Germany, preaching the doctrine of the rebaptism believers, the imminent return of Christ, the "sharing of all

3. *Ibid.*
4. *Ibid.*

things." Led by Thomas Muenzer, the peasantry organized themselves to fight the established church and the nobility. The rag-tag peasant army was soon destroyed, and in 1525 Muenzer was beheaded. The ideas of the movement did not die with Muenzer, however; the remaining members continued to live communally; all money was held in common, and all clothing, bedding, food, and the like were stored in a central depot.

Further, the Anabaptists began to reorganize the family unit itself by incorporating polygamy and divorce into the system. Hence, the movement was truly radicalized, as it challenged the basic assumptions of the rigidly stratified Europe of its day.

The Anabaptist movement died nearly four hundred years ago. Yet its death[5] was not entirely in vain. Groups such as the Mennonites, the Amish, and the Church of the Brethern are its offspring. Perhaps more importantly, the Anabaptist movement to restore pure Christianity has revived in a recent communal experiment in Chicago.

In the 1950s, a group of young Mennonite idealists began to lay the groundwork for a radical, evangelistic Christian community. Unlike most religious utopians, the group decided to stay in the city where they could serve the poor, the alienated, and the disenfranchised. In 1957, John and Louise Miller purchased a home in Chicago. Gradually, their friends and compatriots followed, moving into contiguous areas.

> When possible we want to live within easy walking distance of one another. Scattered as many of us are during the working days across the sprawling network of the city, we want to come home at evening time to one neighborhood where we are readily available to one another in times of need. We want to be able to meet daily if necessary, without climbing into our cars and going half a city away. We want our children to grow up experiencing more than the lonely crowd. We want them to know in their daily life the reality of a closely knit circle of families and friends.[6]

More than 100 men, women, and children belong to the Reba Place Fellowship. They meet several times a week for decision-making, worship, singing, and on Friday evenings, a common meal.

Reba Place communalists take seriously injunctions of Jesus to "go, distribute to the poor." The first-century Christians shared their property

5. For an analysis of why the movement failed, see Belfort Box, *The Rise and Fall of the Anabaptists* (London: Swan Sonneschein, 1903).

6. From a pamphlet, "The Way of Love" (Evanston, Illinois, Reba Place Fellowship).

communally, and Reba Place members point to this as an ideal for the twentieth century.

> The early Christians after pentecost were inspired by this . . . revolutionary attitude when they stopped calling their possessions their own and began sharing freely with each other.
>
> We feel it is urgent that men find this same loving way in economic life today. We want to make a beginning in our fellowships by asking all our members to cultivate that inner detachment from things which will make a just and equitable sharing of goods possible. To facilitate this, we have adopted the old Christian practice of a common treasury to which we bring our economic assets and earnings. From this treasury the members receive a living allowance based on subsistence standards, as well as support for other necessities such as housing and medical expenses. Fellowship funds in excess of what is needed for the support of its various members and guests are then distributed elsewhere.[7]

In fact, the living-allowance is nothing short of spartan for Reba Place communalists; in 1970, a family of five could expect a monthly allowance of $130 for food, $59 for personal spending, and $7 for household spending—a grand total of $196. Medical expenses are paid by the group as a whole. Houses and several cars are collectively owned. If a surplus occurs in the group's treasury, it is almost immediately donated to a charity, an antiwar campaign, or the like. Obviously, the Reba Place Fellowship, like other communal varieties, requires a "scaling down" of the needs of the individual. This is very difficult to accomplish in a society that continually brain washes the individual with the absolute necessity of speedboats, electric toothbrushes, color television, liquor, etc. Individuals can make sacrifices for groups only when groups offer a great deal of emotional support. Common goals and a kind of nonjudgmental openness seem to make Reba Place a highly supportive institution.

Reba Place members make a conscious effort to be honest. They argue that

> there are two areas of human experience where honesty often breaks down, and the lie gains foothold, working havoc. One is at the point of guilt; another is at the point of judgmental attitudes. When we violate our conscience, we experience an internal reac-

7. *Ibid.*

tion not unlike pain. It is the sign that we are moral beings and alive to the reality of God and our fellowmen. This painful realization of sin is guilt. At the point of feeling guilty, we are very susceptible to the lie, for our instinct at that point is to hide. . . . Right here is the source of many an emotional and social disturbance, and there is no answer to it other than the restoration of integrity. The guilt ridden man must acknowledge what he most deeply knows himself to be. He must stop hiding and make a new start with others in genuine honesty.

Just as troublesome as guilt as a potential breeding ground for deception are those judgmental attitudes we often carry around against others. Guilt has to do with our sins. Judgmental attitudes have to do with another's sins. . . . Where my sin is involved, the only escape from deception is confession[8] and forgiveness. Where my brother's sin is involved, the only honest way is loving admonition in a spirit of genuine humility. . . . Both of these we seek to practice in our fellowship.[9]

Reba Place communalists believe that injustices such as hunger, racism, and war must be confronted. They also believe, however, that individuals who confront the larger society must have group support. The nuclear family cannot in many cases bear the brunt of the emotional demands of its members. While Reba Place individuals are monogamous and traditional in their viewpoint toward sex, they feel their communal group provides much more emotional security than the isolated family. "The individual," they point out,

. . . once nurtured in a stable, larger family or community, is making more and more decisions alone, unsupported or unchallenged by anyone else. There is no one else. . . . The individual family was never meant to carry the load it is now trying to carry. The small family of mother, father, and children needs a larger supportive context. It thrives best in the give and take of a closely knit community of families.[10]

Indeed, the families at Reba Place do seem to thrive. Members work with retarded children, run a coffee house, do social work, teach, and work with the emotionally-disturbed.

8. The group confession was a practice common to many nineteenth-century communalists.

9. Reba Place Fellowship, *op. cit.*

10. *Ibid.*

KOINONIA FARM

The Greek word *koinonia* means fellowship or community. In this case, Koinonia is a biracial fellowship located near Americus, Georgia. The American South has not been noted for its openness toward utopian movements, let alone racially-integrated ones. The attitudes of many rural Georgians toward Koinonia Farm is no exception to this dictum.

In 1942, Clarence Jordan founded Koinonia with the purpose of bringing a radical form of Christian living to a typical deep South county. The radical part of Jordan's message lay in his idea of a communal partnership of blacks and poor whites, farming together as equals, indeed, as brothers. With very little capital and great determination, the farm began to grow and, to some extent, prosper. Koinonia Farm is organized on the basis of "partnership farming." This means that each partner is given a loan (*e.g.*, land tools, and a house) to produce a cash-crop of corn, pecans, cattle, or peanuts. The surplus profits from the sale of the crops goes to a common fund, which begins new ventures and aids the communalists' charitable projects.

The farmers vary in education from local blacks, who are barely literate, to Clarence Jordan, who held a Ph.D. in theology. The ages of Koinonia communalists range from teenagers to those in their seventies. In the summer, students from Eastern colleges work shelling pecans or doing farm labor. Other young people labor at the farm as part of their alternative service to the draft. Several Mennonite youths work full time at the farm, and one young Canadian girl from a Hutterite Colony lives at Koinonia while attending a nearby all black high school in Americus. Other Koinonia partners have close ties with the Reba Place Fellowship.

Clarence Jordan, the real moving force of the farm, died October 29, 1969. His influence and ideas are still felt at Koinonia, however. It was Jordan himself who felt the direct sting of persecution by the local Ku Klux Klan. In 1957, Jordan found the Koinonia Farm market dynamited. Also in the midfifties, the White Citizens' Council formed an airtight economic boycott of the farm. Passing cars would shoot into the house, and communalists were often beaten as they went into town. Still, Jordan's leadership and the efforts of the partners prevailed. Koinonia now has 1,400 acres of land. The housing program completed four new houses in 1969–70.

Shortly before his death, Clarence Jordan wrote of wishing to acquire 5,000 more acres of land. Chief among his concerns had always been the rural refugees in the South, who were rapidly losing their meager jobs to the automated machinery of large commercial farms.

> Partnership housing is concerned with the idea that the urban ghetto is to a considerable extent the product of rural displacement. People don't move to the city unless life in the country has become intolerable. They do not voluntarily choose the degrading life in the big city slums; it is forced upon them. If land in the country is made available to them on which to build a decent house, and if they can get jobs nearby to support their families, they'll stay put.[11]

Death put an end to Jordan's dreams. Yet, other members of the commune were prepared to carry on. The group maintains an extensive mailing list to contact sympathetic sources for financial support. Through their sales of candied pecans, fruitcakes, books, and records, the communalists are able to bring much needed cash into their treasury. One of the groups most popular items is "The Cotten Patch Version" of several books of the Bible. The Cotton Patch Version, as written by Clarence Jordan, attempts to translate the good book into a language and imagery meaningful to rural Southerners. Jordan conceives of St. Paul as a converted Southerner who confronts the problems of alienation and racism with brotherhood and Christian living.

With a little help from its friends, Koinonia is likely to endure. Its goals, as outlined by its founder, are three in number. First and foremost is

> *Communication*—The sowing of the seed, the spreading of the radical ideas of the gospel message. . . . It means to preach good news to the poor, to proclaim release to the captives and recovering of sight to the blind, to set at liberty those who are oppressed, to proclaim the acceptable year of the Lord.

The second goal is

> *Instruction*—The constant teaching and training of the partners and those who seriously want a new way of life built around the will of God, to enable them to become more effective and mature.

11. Pamphlet.

> There will be traveling "discipleship schools" to follow up and conserve the results of the speaking and communicating. . . .
>
> The final goal of the Koinonia movement is *Application,* which
>
> consists of partnership industries, partnership farming, and partnership housing. These will be implemented through a fund for humanity. . . .

Although these dreams are modest enough, the communalists realize their task will not be easy. Dedicated as they are to racial equality, the Koinonia partners face great frustrations in their environs. During the "reign of terror" endured by the farm in 1956, most blacks left the farm for their own survival. By 1970, many blacks avoided the farm for another reason; the "white man's religion," Christianity, was out of favor with militants. The faithful at Koinonia are not likely to give up, however.

Two days before his death in 1969, Clarence Jordan wrote

> . . . we have sought out dedicated people to share the dream and to be our partners. Materialism, competitiveness, and self-interest are so deeply entrenched in our culture that they have almost exterminated the spirit of partnership and sharing. But people with this spirit *have* been coming. Many more are needed, but we have faith that, in time, they *will* come. The breed is not yet extinct.[12]

QUAKER EXPERIMENTS

The Society of Friends has long been known for its social concerns and nondogmatic theological position. Quakers were in the forefront of the Abolitionist movements and movements for prison-reform. Moreover, predominantly-Quaker towns have had a reputation for charitable works and community-spirit. More recently a number of members of the Society of Friends have concerned themselves with the creation of "intentional communities." The movement for intentional communities is small at the present, as is the Society of Friends. Yet, the movement's importance is likely to have far more significance than the number of its participants would imply.

The Vale. Yellow Springs, Ohio has less than 4,000 people within

12. *Koinonia Partners' Newsletter,* October 27, 1969.

its boundaries and is officially classified as a village. Nevertheless, it is a most unusual village. It is, among other things, the home of the American Humanist Society, an organization dedicated to conflict with supernaturalism and organized religion. Moreover, it is the home of Antioch College, an institution known for its academic excellence and for its socially-committed students.

A few miles south of the village lies thirty acres of gently-rolling hills and rather dense stands of shrubbery and trees. Less than a dozen houses, including a geodesic dome, are scattered throughout the forest. A drive through the graveled roads in the evening would likely find neighbors playing volleyball or working in the fresh air. This is the Vale, an intentional community founded around 1960 by a group of Quakers and their friends. By 1970, five families were full members of the communal venture. Several more have rented from the community and are sympathetic to the group's goals. The Vale, like Koinonia and the Reba Fellowship, is not simply a retreat from the larger society. Nor is it the sharing "of all with all." It is, foremost, a group of people involved in mutual aid and common concerns. The thirty acres of land outside Yellow Springs is owned by Vale members in a mutual "land trust." The private homes of the members are far enough apart to provide a maximum of privacy with a minimum loss of community spirit. Beyond the common ownership of the land, Vale members share in other practical ways. For example, Vale communalists and their neighbors buy health foods collectively at a wholesale rate.

More importantly, the Vale members share the emotional support of those involved in common endeavor. One member of the group is Dean of Students at Antioch. Another, a widowed mother of four, is leader of the Yellow Springs Youth Orchestra. A third is a printer and co-publisher of the socialist-pacifist *Yellow Springs News.* Two of the group spent terms in prison for draft refusal, and most other members of the Vale have been active in antiwar and American Friends Service committee work.

Griscom Morgan, one of the driving forces in the organization of the Vale, has concerned himself with the intentional communities movement for more than twenty years. His father, Arthur E. Morgan, was formerly president of Antioch College, chairman of the Tennessee Valley Authority, and the first president of Community Service Incorporated, a research and publishing company devoted to the study of the small community.

Griscom Morgan has spent much of his adult life studying intentional communities such as those begun by the Seventh Day Adventists in the southeastern United States. Currently, Morgan's Community Service Incorporated has concerned itself with a number of widely diverse community experiments, such as a Quaker-oriented commune in Ghana, the Mitraniketan Community in India, the Celo Community in North Carolina, and the Seneca Nation Community in New York State. Community Service's latest concern was with the Hopi Indians of Arizona.

The Vale and its sister institution, Community Service, act as a kind of switchboard for the communal experiments in the United States. Recently, for example, the Hog Farmers stopped by the Vale on their way to New York State. Griscom Morgan was impressed, he told me, with the perceptiveness and seriousness of the busload of Hog Farmers. "They may have left the others because they were more serious and perceptive than those they left," he commented.

Drugs are not allowed at the Vale. Even smoking is proscribed. Griscom Morgan believes that drugs are not only potentially dangerous to the individual, but also a destructive force to the establishment of communities. "That is why working-class people frown on drug use so much. They realize they have to stick together to survive. They can't have solidarity with drugs. They know that. A lot of middle-class kids haven't found that out yet." Morgan went on to compare the use of drugs in a communal setting to jumping up on to a large wall and seeing a beautiful garden. "You get a glimpse of the beauty, then you come down hard with the realization that you can't have the beauty in your day-to-day life." Griscom Morgan's conversations with hip communalists have convinced him that drugs "privatize" or "individuate" the natural leaders of communal groups to the point that they no longer care to keep order or resolve group conflicts.

Morgan believes war and exploitation will end only when individuals plan their lives in a creative and nondestructive way. "To do away with war in the abstract is not our business." He argues,

> Our responsibility is to accomplish what is within our power in creating a just and progressive social order free from the seeds of destruction and violence. The old order is in the process of destroying itself. We must build a viable alternative to supersede it. . . . The opportunities at hand are almost untouched for us to labor in displacing the sources of disaster threatening mankind by building an alternative social order through living it with our own lives.

> Heretofore most people have been fighting the old, unjust social order while enmeshed in it and supporting it *by being part of it.* If we will begin with ourselves and then join in association with others who are of like mind in building a better social order, the widening ring of effect and growth will be progressive. This means that after ourselves and our families, the fellowship, the small community, and the fellowship of such communities are our opportunities to create a new and progressive social order that will bring world peace and social justice.[13]

Morgan believes, as do most communalists, that society is too large. The best argument for communal living, he feels, is the fact that man does not truly adapt to the overcrowding of modern city life.

> It should be clear that there is no such thing as healthy acclimation to such effects of crowding, any more than there is an acclimation to atomic radiation from exposure to it. Brief periods of radiation or of crowding may be unharmful if, on balance, there is a relative freedom from it. Seals, birds, and buffalo all had their times of herd crowding, but these were balanced by long periods of isolation.[14]

Morgan goes on to make his case for a smaller social unit. Overcrowding, he believes, causes "overstimulation."[15]

> Whether overstimulation from overcrowding takes place through physical principles and influences, we have yet to discover, or through those we already know, remains to be determined. But certainly overstimulation does exist, and we do not yet know how it works. There are no sound grounds for assuming that we can overcome and compensate for harmful effects of overcrowding . . . when we are in ignorance of how these effects take place. The number and frequency of personal interactions is not the primary cause of the harmful effect of crowding. Many a villager has far more social interactions than the average urbanite; yet, it is the

13. Griscom Morgan, *Intentional Community Handbook* (Yellow Springs, Ohio: Community Services Inc.), p. 2.

14. Griscom Morgan, "The Human Scale in Schools," *Community Comments* (June, 1970): 8–9.

15. Morgan's idea of overstimulation is analogous to Marshall McLuhan's concept of "sensory overload."

> urbanite that suffers stress from overstimulation from his environment.[16]

Jane Morgan, Griscom's wife, teaches at a school (grades 1, 2, and 3) owned by the Vale. It goes without saying that the primary school is not overcrowded.

Members of the Vale are thoughtful, unassuming people. Yet, it is their belief that the establishment of one successful community may have transcendent significance.

> A mature community can send off, or swarm off, like bees, an offshoot group to begin a new development based on the background, competence, and experience of the older community. Thus many small communities have been established by alumni of the Adventist community at Madison, Tennessee, following the pattern of the parent community. Most successful pioneering community ventures have had members who have had experience in some significant older community.[17]

The Bhoodan Center. The word *Bhoodan* is Hindi and was used by Ghandi to denote the sharing of land. The Bhoodan Center of Inquiry is a Quaker-oriented research bureau on communal living. It is located near Oakhurst, California and is one of the few religious communes in that area that has survived. From 1933 to 1960, intentional communities were organized in Modesto, Temple City, San Fernando, Grass Valley, and other California cities.[18] Most failed.

Members of the Society of Friends at Oakhurst were not discouraged by the failures of like-minded groups. They began a two-pronged attack on the problem. First they began to establish a permanent center for the study of communitarian problems. The center was to take no dogmatic positions on social issues or communal doctrines but was to serve as a clearing house for dialogues on the nature of community and the practical problems of formulating an intentional community.

Past communities have failed for essentially three reasons, according to the Bhoodan communalists. First is the mistaken idea that a shared piece of land constitutes a community. "People who come to us and

16. Griscom Morgan, *op. cit.*
17. Griscom Morgan, *Intentional Community Handbook*, *op. cit.*, p. 5.
18. See *Manahar Cooperative Fellowship Handbook* (Oakhurst, California: Bhoodan Center, 1970).

say, 'I want to join your community,' should rather say, 'May I come and help build a community with you?' Few come with the realization that they will only have what they build. Most come expecting to join a finished product."[19]

Second,

> a community should not be based on an economic enterprise. If the enterprise fails, and this is all the group had in common, it would mean they would separate when they needed each other the most. Community should rather be based on the agreement of a group of people that they are going to help one another work on any and all of their problems.[20]

Finally, the Bhoodan group sees another error in the use of majority rule by communal organizations. This is perhaps due to the group's Quaker background. Traditional Friends' meetings have attempted to gain a "sense of the meeting" or unanimity in decision making.

> . . . we strive for unanimity, even though it takes longer. When unanimity is not achieved, we try to find ways that those differing can follow their various initiatives in their different ways while continuing to seek direction and cooperating with the group in other things.[21]

The Bhoodan Center led by its president, Charles Davis, has engaged in a number of projects relevant to the development of intentional communities. One experiment, conducted with the aid of Antioch College students, was the construction of a residential dwelling by the rammed earth technique. This technique, which involves very little capital outlay, requires the mixture of cement and damp soil (in a twelve to one ratio). The mixture is tamped into a form. The form is then moved to the next section. In a short time an entire dwelling can be erected in this way. The method has obvious possibilities for the rapid construction of low-cost housing.

The Bhoodan Center also sponsors a number of work-study seminars for young people. The students study social problems and community

19. *Ibid.*, p. 3.
20. *Ibid.*, p. 4.
21. *Ibid.*

development, and are greatly interested in the culture and social organization of the Hopi Indians. In June of 1970, a dozen from the center visited the Hopis with the idea that they could learn from "the Hopi way" how to deal with the problems of small community development.

If the Bhoodan Center is the theoretical and experimental side of the Quaker utopists, the Manahar[22] Cooperative Fellowship is the practical application of their beliefs. The Manahar community is about 80 acres of rather mountainous land near Oakhurst. Charles Davis writes,

> Those of us here now are not in favor of strictly communal living, with everyone living together in one residence and interacting with each other day and night and sharing everything. We share many things, but on a voluntary, not a compulsory, basis. We have developed a closeness among us and are learning a great deal about human relations by experience, but, although it is necessary now with our limited facilities, we do not plan to always live so physically close nor to always eat communally.[23]

Members of the fellowship do share many things, however, including a two ton truck, a Volkswagen bus, a Land Rover, tools and the like. The Manahar people also buy collectively and exchange labor with each other. Currently, some of the community's members are attempting to grow crops in the Sierra Nevada foothills.

The Manahar-Bhoodan experiments are not primarily economic in nature. Basically, the members cooperate with each other not because of any financial theory, but because they share common concerns. These concerns are primarily religious; although they are not doctrinaire or dogmatic. (The communalists see religion as a search rather than a set of beliefs.)

CATHOLIC WORKERS—TIVOLI FARM

It would be a mistake to regard the Roman Catholic church as a bastion of conservatism. The church is, of course, tied to tradition; but the tradition of the church involves radicals as well as reactionaries. The nexus of Catholic radicalism in the United States has been the Catholic Workers' Movement. The movement had its beginnings in

22. The Hindi word *Manahar* means "harmony of man and nature."
23. Manahar, *op. cit.*, p. 8.

1932,[24] when Dorothy Day, an excommunist turned Catholic, met Peter Maurin, a French emigré who had come to America with doctrines of radical Catholicism and anarchism. The worker movement began in earnest to better the lot of the poor. Scraping together a few dollars, Dorothy Day and friends opened a Hospitality House for the sick, the homeless, and the jobless in depression-torn New York City. "Inasmuch as ye have done it unto the least of these, ye have done it unto me." The "least of these" were plentiful in the city, and the workers began to provide housing for them in a semiheated building in the Bowery. A few years later they started a garden commune near the city. In the years that followed, Dorothy Day and her co-workers became involved in numerous service projects and social experiments. Michael Harrington, author of *The Other America*[25] and Congressman from Massachusetts, worked several years in the city with them. The Workers' Movement produced other deeply committed radicals as well. Ammon Hennacy, who died in 1970, was an anarchist-pacifist involved in the movement for several decades. Hennacy refused to pay income taxes "to feed the war machine." During the last few years of his life, he moved from New York to Salt Lake City, where he founded Joe Hill House.[26] "It is too far from the tracks for transients to find," Hennacy complained, only a few months before his death. "Only a dozen can be accommodated, and it is meant for students and young folks who come and go. I attend meetings there every Friday night."[27] Hennacy, who went to Utah to promote pacifism, communal living, and aid to the poor, also wanted to confront the Jim Crowism evident in the Mormon city.

The base of Catholic Workers' activities remains in New York, however. They maintain a communal home for societal outcasts at 36 East First Street in the city, along with Tivoli, a communal service farm in upstate New York.

"The homeless, the jobless, the littered continue the retinue view to our door," writes Pat Jordan, a worker in the Bowery.

> Several people spoke of blood banking as their only sure source of revenue. Swede and Manny approached different hospitals with

24. For a history of the early movement, see Dorothy Day, *House of Hospitality* (New York: Sheed and Ward, 1939).
25. (New York: Penguin Books, 1962).
26. Named, of course, after the famous wobbly radical of the 1920s.
27. Ammon Hennacy, *The Catholic Worker*, January, 1970, p. 5.

similarly lacerated heads. A Puerto Rican man perished near the bacci courts at the corner from knife wounds. Hillies new bar on the Bowery brazenly locked out and outlawed "all Bowery derelicts" to serve only the artsy. An old black woman fell asleep on our first floor with no other roof and no other bed.[28]

Days are not always depressing at the Workers' communal house for

. . . even yet in the Second St. litter of this May, an old liner found a copy of Nietzche, and his gleeful eyes said this find was like rediscovering nature. At Cooper Square, the frisbee saucers came arching out, and cars had to slow down for the avenue of gliding fun. Polish Mike brought flowers to the girls. Carol Hinchen and Mike Scahill returned radiant from Cuba. Arthur Lacey sported bright orange trousers. And some of our friends said they'd try to condition their drinking to one day at a time. . . . Outside an anarchist paint-in has left the front mid-way between the blue meanies and the yellow submarine. . . . All this newness is a reminder to seek life, reject illusion, and cut thin ones patness.[29]

Ron Gessner, manager of the Tivoli collective farm, writes,

Our community is a bit difficult to describe. We have been in existence for over 30 years and have been many different things in past times. At present, we are a group of some 50 persons, in age from two months to 70 years. We are basically a community of need. Some of us are sick, some old, some ex-alcoholics, and some volunteers.[30]

The day-to-day problems at Tivoli concern plans for building a new chicken coop, repair of the badly leaking main roof, finding a serviceable car to traverse the bad roads, and, of course, the solicitation of money and volunteers to do the service work. The workers are generally not greatly concerned about political or social theory. They have opted for a decentralized community that emphasizes in a practical way their concepts of Christian love. Like the Quakers, they have a strong "peace testimony," which is manifest in their antiwar efforts as well as their service in "the community of need."

28. Pat Jordan, *The Catholic Worker*, June, 1970, p. 2.
29. *Ibid.*
30. From a letter to the author.

THE RELIGIOUS FACTOR IN COMMUNAL LIFE

Communal groups such as the Lama Foundation or the Brotherhood of the Spirit, a mystical, E.S.P.-oriented community in Massachusetts, are in many ways different from those built on Christian traditions. One concern they do share is the need for transcendent values. That is to say that, unlike many hip drop-in communes, religious communes share a concern beyond their immediate needs. Whether their needs be some form of disciplined self-realization or an altruistic goal, religious communalists share the idea that their work can be an example to others in the mundane world. Some Christian communalists point to the biblical idea of a "saving remnant," which is "called" to set an example to their fellowmen. Hence, living an exemplary life, based on cooperation and lack of material concern, can show a standard to the rest of the world.

Milton Rokeach, in an article in the *Review of Religious Research*, cites his findings that those who hold orthodox Christian theological views are unlikely to exhibit attitudes of social compassion.[31] In fact, one of Rokeach's findings indicated that an individual's belief in salvation was negatively related to a compassionate view of events, such as the killing of Dr. Martin Luther King. Curiously enough, those church-goers who stressed the idea of forgiveness in their theology were no more likely to show social compassion than were non-church-goers.

The otherworldly attitude of Christianity has been attacked frequently by humanists who maintain that the concern for salvation betrays Christians' failure to deal with social realities and problems. This criticism cannot be leveled at the radical Christian communalists described in this book. In varying ways, the radical Protestants and Catholics living communally are confronting the problems of twentieth-century America. In no sense do the radical Christian communes proximate the escapist monastic communities of prior centuries, for the radical Christians believe that isolation can maintain religious purity only at the expense of concern for ones' brother. Older groups of Chris-

31. "Religious Values and Social Compassion," *Review of Religious Research* 2, no. 1 (Fall, 1969): 24–37. Rokeach defines social compassion as concern for the poor, the downtrodden, or minorities. One illuminating finding by Rokeach was that about one-third of the 1,400 Americans sampled felt that Dr. King "brought it [the assassination] on himself."

tian utopists attempted to perfect society by escaping it.[32] Modern Christian utopists serve the larger society while deriving their strength from the development of intentional communities.

In general, religious communes are stabler and longer-lasting than hip communes. As I have indicated, men need a sense of purpose beyond their immediate needs. This might be called a need for transcendence, or simply, an ideology. Whatever it is called, religious structures fulfill the sense of purpose for vast numbers of people in many varieties of social organization. Many hip communes begin with an absence of goals or purpose; hence, hip communalists often lack the *esprit de corps* to maintain their communities. Religious groups, on the other hand, begin with an advantage in community-building because they share a common ideology. The problem with most religionists is that they resist social innovations of almost any variety. The *status quo* takes on the appearance of the sacred for most traditional believers. Thus, paradoxically those most likely to succeed in new community-building are those least likely to try.

Christian communes such as Reba Fellowship, Buderhof Society, Bhoodan Center, and Tivoli Farm are not doctrinaire in theology. Hip communes such as Morningstar and Wheeler Ranch are currently developing religious tenets. Some day the movements may converge.

32. A good case in point is the nineteenth-century Mormon church, which moved with the American frontier to escape the world with the purpose of building Zion, the kingdom of God on earth.

7

RADICAL COLLECTIVES AND WOMEN'S COMMUNES

"Our work is guided by the sense that we may be the last generation in the experiment with living."

from the Port Huron Statement

"Sisterhood is powerful."

a slogan of the Women's Liberation Movement

URBAN REVOLUTIONARIES

The ideas of decentralization and mutual cooperation antedate the founding of the American republic by hundreds of years. Peasant communes have been common in many areas of Europe.[1] During revolutionary times, radical, underground communal groups are common to both Europe and Latin America. Some claim that there is a revolution in the United States now. Whether this is correct or not, some individuals, fearing political repression, have banded together in communes. Unlike hip communalists, most radical communalists have no desire to get back to the land. For the most part, they are "where the action is," and this means the city. A number of "resistance" communes have sprung up as antiwar advocates. They are convinced that group solidarity is a necessary component of a protracted struggle against the Vietnamese war. Some of these groups have sprung from traditionally pacifistic groups such as the Quakers. Resistance House and Any Day Now, in Philadelphia, are good examples of this variety. Individuals in communes such as these live together informally as they participate in

1. Especially Spain and Italy; See E. J. Hobsbawm, *Primitive Rebels: Studies in Archaic Forms of Social Movement in the Nineteenth and Twentieth Centuries* (New York: Frederick A. Praeger, 1963).

draft counseling and other antiwar activity. For the most part, they do work outside the communes to provide for their existence and for the furtherance of their political causes.

Other far more radical or even revolutionary groups have developed underground communes as well. The Weathermen faction of S.D.S., which has been accused of bombings and other terroristic activities, has formed underground "collectives" which provide mutual protection[2] and support for the young revolutionaries. Weathermen collectives or underground communes have been said to provide an effective network of escape for political refugees. Dr. Timothy Leary, the guru of the quasipolitical psychedelic religion in America, escaped from a minimum-security prison in California in the late summer of 1970. Weathermen were quick to take credit for Leary's escape.

The number of resistance communes is actually small compared to the "drop-out" variety. Furthermore, they are difficult to locate due to their desire for secrecy and their fear of infiltration. It does seem to be true, however, that communal revolutionaries such as the Weathermen or the White Panthers[3] are "into the cultural revolutionary thing." That is to say that they have come to believe that the development of a revolutionary life-style is fundamental to the political revolution. Hence, the radical communalists smoke dope and groove on hard rock music and freaky appearance.

In 1970, one of the most radical underground newspapers in southern California, *Contempt*,[4] endorsed the communal movement and pleaded for more militancy and political awareness for all communal dwellers. The fact is that this is now unlikely, since many communalists have recently left the radical political scene.

In 1969, in Cincinnati, the Weathermen collective was especially paranoid about the possibility of police or F.B.I. informers. One new member was given the "acid test" (L.S.D.) to determine his sincerity about the revolutionary group. The new member passed the test rather well by not saying or doing anything counterrevolutionary during his trip. He was, however, having trouble with his sex life. A young Weatherwoman formed a liaison with him to work out his sex hangups.[5] All

2. Part of the protection is sought through the study and practice of karate.
3. The White Panthers formed a commune in Ann Arbor, Michigan in 1970.
4. "Communal Life Means Struggle," *Contempt* 1, no. 3, p. 13.
5. A visitor to a Weathermen collective confided to me that women in the group were "available" to any male revolutionary "on call." The reverse was not true. This, if true, leaves this Weathermen group open to the charge of male chauvinism.

was going well until some of the young revolutionaries noticed their compatriot coming out of a Federal building in the city. Later, two Weathermen were arrested, and rumors circulated around the Cincinnati collectives that a "pig" or informer was in the group. "L," the accused informer, still maintained that he was a true revolutionary. Yet after the group was "busted" for its activities, "L" alone was given reasonable bail—$7,500 as opposed to $20,000 to $50,000 for the rest. At this point "L" was thrown out of the collective. His revolutionary girlfriend was pregnant, sick, and severely distressed at the thought that her lover might indeed be a counterrevolutionary. "L" moved on to Berkeley, where he tried to involve himself in revolutionary activities. The underground press,[6] however, reported "L's" activities with the result that he was rejected and "put down" by nearly all the urban leftists. "L's" case is not unusual, and it demonstrates why collectives such as those of the Weathermen are so secretive. "We all know what should happen to 'L'," comments one revolutionary[7] ominously. Revolutionaries, if they are truly committed to violent overthrow of a political system, are prone to use violence in what they regard as their own self-defense. At any rate, Weathermen are nearly totally alienated from the larger society. They are at war with the government, and the collectives are the battalions of the new urban guerillas.

The Weathermen are the most controversial, indeed, notorious, of the revolutionary collectives. Yet, they are not the only political organizations opting for urban collectives and communes. Other less publicized groups also use the collective life-style as a political tool. Some are local organizations whose members left S.D.S. or other organizations because of their disagreement with extreme dogmas.

A case in point is the Seattle Liberation Front.[8] On April 16, 1970, eight members of that group were charged with "conspiracy and crossing state lines to incite a riot." This stemmed from massive demonstrations in Seattle protesting the trial of the famed Chicago Eight. The group had organized four collectives. The four multiplied by dividing, and by 1970, about twenty collective groups were formed in working-class neighborhoods of Seattle. Boeing Aircraft, the ninth largest war contractor in the world, was in the process of cutting back on its work force. The young revolutionaries hoped to translate worker discontent

6. See *Liberated Guardian*, September 27, 1970, p. 7.
7. *Ibid.*
8. For an account of the group, see Michael Lerner, "Conspiracy in Seattle," *Liberation*, July, 1970, p. 36.

due to the layoffs into revolutionary fervor. "We felt all along," said one member of the Front,[9]

> that collectives were the correct form because they allowed people in some small way to incorporate a version of socialism into their daily life, which is very important. . . . We're developing different collectives into all sorts of different things. We started with what you might call serve the people programs which we feel are very essential. We had a program at the food stamp offices, where there's 500 people every day at seven in the morning waiting outside till it opens at eight. . . . So we started a breakfast program there where we feed them hot breakfast—orange juice and coffee, etc. We had different collectives doing that everyday. . . . We also had a street collective of street people. They were the first people to start a food program, totally on their own. They would feed anywhere from 80 to 120 people hot dinners, really good spaghetti dinners. . . . This summer the street collective is going to develop some approach to smack [heroin]. . . . We feel that the only way to beat smack is with some sort of meaningful alternative for kids to get involved in. The reason people take to smack in the first place is because they're so fucking alienated from the system, and you can't tell them to rejoin the system. The revolution must have forms for people to get into. We think a collective form is one way.

Groups like the Seattle Liberation Front see communal living in the urban scene as a source of potential revolutionary strength. Moreover, they reject the mindless "trashing" and terroristic tactics of ultra-Leftist Weathermen. Members of the Seattle collective also seek to capitalize on local issues, such as unemployment or the potential displacement of low-income people from their homes by freeway I-90. S.L.F. organizers are quick to remind us that they are not reformist in their goals.

> Precisely because our task is not only to destroy capitalism but also to radically remake ourselves, the present historical period calls for organization built around collectives of 10 to 15 people. It is only in collectives that we can develop ourselves as creative political organizers without the stifling atmosphere that the large mass-meeting based organizations like S.D.S. made current. In the

9. "Seattle Liberation Front," *Vocations for Social Change*, September, 1970, pp. 33–34.

difficult struggle to transform the movement from the male-dominated, easy going, non-ideological, and anti-intellectual fun and games of the sixties into a tough and sensitive grouping prepared to withstand the repression and reach working class constituencies in the seventies, the collective form will be crucial. The collective forms allow us to build trust, mutual love and struggle, and will liberate the creativity and imagination which must still be among our chief weapons.[10]

The urban collective is a new, comparatively small movement within the repertoire of the *soi-disant* revolutionaries. It provides togetherness and a life-style that resists co-optation by the larger society. Togetherness may have its disadvantages as well as its assets for the revolutionary movement, nonetheless. If the movement is suppressed by legal or even extralegal attack, the revolutionaries will be in one place, thus making it easy for authorities or vigilantes to destroy or cripple the revolutionary organization.

WOMEN'S COMMUNES

One of the movements that most directly had its genesis in the early civil rights movement was the movement for equal rights for women. In the two-year period between 1968 and 1970, books, magazines, pamphlets, and newspapers all focused on the dilemmas of women. A number of radical young women came to perceive the "sexist" social system at fault for the dehumanizing position of women who feel compelled to cover up their intellect, sell their bodies, turn down career opportunities, and dedicate their lives to washing, cooking, ironing, and generally pacifying the ego of the male animal. Many women in the movement have come to see their position in the society as analogous to that of the black man. Both have been denied the right to vote, both have been discriminated against in terms of jobs,[11] and both have been dehumanized with derogatory names.[12]

Consciously then, women in the movement began to pattern their

10. From a Seattle Liberation Front organizing paper.

11. For an analysis of wage and job discrimination against women, see Marlene Dixon, "Why Women's Liberation?" *Divided We Stand* (San Francisco: Canfield Press, 1970), pp. 32–38.

12. To many women the idea of being called "a chick," "a bird," "a piece," or "a dog" is as offensive as calling a black man "boy," "a buck," or "nigger."

propaganda,[13] and to a lesser extent their organizations, after the civil rights and black power movements. It is important to remember that the women's movements, like the movements for racial equality, vary in ideology, technique, and degree of moderation. Some groups such as Betty Friedan's National Organization of Women ("N.O.W.") are moderate and perhaps comparable to old-line civil rights groups such as the N.A.A.C.P. or the National Urban League. Some groups of a more militant variety, such as the Women's Liberation Front, are comparable (in rhetoric, at any rate) to revolutionary groups such as the Black Panthers. Significantly, the movement for black equality has always produced a minority of advocates of separatism, such as Marcus Garvey's "Back to Africa" movement or current Black Nationalists, such as the Black Muslims. Thomas Pettigrew, a social psychologist at Harvard, calls this alternative "moving away from the oppressor."[14] Pettigrew maintains that one option open to oppressed people is that they can reduce their frustrations and tensions by seeking to avoid contacts with those who pose a threat to their self-concept or sense of self-worth.

It may seem absurd for women, now, to demand a separate nation. Yet some women in the movement are opting for separation from what they deem the oppressive world of men's domination. How is this to be accomplished? By the establishment of communes—communes established by and for women only.

One such communal experiment for "liberationist" women was established in 1970 near the University of Iowa.[15] "We have lived together in a large house for just about one month," writes one of the women.

> There are six sisters who actually live in the house and two more sisters who will join us in the fall when they return from traveling. Two of the sisters help organize and run the two free day care centers in Iowa City. Other sisters in the house help at the day care centers and work full-time at straight jobs. All the money that people earn is pooled together. Sisters contribute anything from a full month's salary to food stamps only. This all goes into either our household kitty or our checking account.

13. There are many equivalent words in the black power and women power movements. "Racism's" equivalent for women is "sexism"; "skin privileges" are "cock privileges"; "white racism" is "male chauvinism," etc.

14. See Pettigrew's *A Profile of the Negro American* (Princeton, N.J.: D. Van Nostrand Company, 1964), pp. 27–58.

15. See "Women's Collective," *Everywoman*, July 31, 1970, p. 11.

Thus far the collective has made several structural decisions. First, we have regular closed meetings to discuss issues or problems at hand. This is a time when we have been consciousness raising. At first we thought that having closed meetings may have been an elitist thing to do. But in the interest of experimenting with a small core of the same people, we decided it was necessary. Also, we have decided that there will be no men in the house. This decision was made in the effort to support the goals that we have made for ourselves. If women are to truly get themselves together, then they need someplace where they may go and always know that the open arms of a sister await them. We think of our house as a place where married sisters, high school sisters, and all others may come to talk if they need to get away for a while. It has been difficult to not allow men in the house, because there always seem to be those exceptions. However, the rule of never any men in our house has established an atmosphere to the house that we all agree is wonderful. It is very important that we do not get co-opted on this point.[16]

The idea of a separatist women's commune is, of course, threatening to the male ego. For this reason, advocates of this variety of communal experiment are likely to incur the wrath and ridicule of most males and a number of women. The Iowa City feminists are not unaware of this; they simply do not consider it important.

. . . In a women's collective, women would learn to look to their sisters, not men, for approval. This will come about through living together, which necessitates decisions and actions that will reinforce strength and confidence.

If the idea of women's collectives sounds threatening, it is—to men. Women are making a definite political move which puts pressure on the men to overcome their male chauvinism. Women are saying our oppression will not be passed over in the revolution as it has everywhere else. . . . Learning to stop competing with our sisters and to start relying upon them will be important lessons in the commune. Once we trust, rely upon, and love each other, it will give us faith and strength to convince other women . . . and to stop males from oppressing us. We'd love and receive

16. *Ibid.*

strength from each other in our struggle to effect a revolution which must liberate women in order to be successful.[17]

While a "minority of the minority" of the women in the liberationist movement formulate separatist solutions[18] to problems of sexual discrimination, others opt for milder means of developing a sense of sisterhood. Perhaps a majority of women's liberation groups continue to hold sensitivity sessions to encourage mutual emotional involvement and a sense of group solidarity.

The movement for intentional communities cannot afford to ignore the contemporary movement for women's rights. Often hip communes, with their prophecies of freedom for the individual, have fallen into the same division of labor as that of the larger society. The women cook, wash, or do other "womanly" things, while the men plant, work in the fields or gardens, and, generally, do "manly" things. One farm commune described by Kit Leder[19] followed this exact pattern. It soon became evident to the women in the commune that they were not liberated in any real sense. The women, who ranged in age from fifteen to twenty-four, began to hold private "caucuses" in the woods to air their gripes about the inequitable situation. This, of course, puzzled and enraged the male communalists, who saw the women's private sessions as the beginnings of a fifth column in their midst. For a time the women went on the offensive. They refused to listen to the harassment thrown at them by their male counterparts. Instead, they began to do "male only" tasks such as chopping wood, driving the tractor, or working in the fields. For several weeks the commune broke into two camps —male and female. A separate "woman's brigade" worked together in the fields.

For a while the women reveled in their new found confidence and skills in providing for their own needs. Learning a simple skill such as the use of an axe became a great source of pride to the women. Nonetheless, in the end some women compromised their idealism and returned to the fold. Others left communal life completely. In any case, the separatist commune failed. The message of this experience is clear, however. The act of pulling away from the larger society does not rid

17. *Ibid.*
18. Most men (including the author) continue to hope for an "integrationist" answer.
19. "Women in Communes," *Women: A Journal of Liberation,* 1, no. 1 (Fall, 1969): 34–36.

a group of that society's influence. If women in the straight society are denied options that would enrich their lives, are they of necessity free in a communal society? The answer is obviously "no."

Yet small-scale, subsistence-level communes may offer a real opportunity for both men and women to "do their own thing," since they do not require highly specialized labor. Ideally, any member of the group could learn to chop wood, plant, hoe, or prepare food. Ideally that is, if the communalists did not bring a rigid (though sometimes unconscious) idea of the proper kind of work to be done by one or the other sex.

The message of the women's liberation movement is that women will no longer be regarded as beasts of burden or sex objects. Women liberationists believe that the present society does not fulfill the human potentialities of its members. Communalists in the main agree. Yet communal living will continue to be subjected to the scrutiny of the women's movement. It is safe to surmise that male chauvinists, even those hiding behind long hair, will increasingly hear of their shortcomings. All of this reinforces a basic idea of this book, which is that one can escape to communal living, but one can not escape the essential question there—how shall I relate to my fellow man and (of at least equal importance) my liberated sisters?

8

WALDEN TWO AND BEYOND

"The Revolution is over: we won!"

title of a pamphlet published by the Twin Oaks Commune

In 1948, B. F. Skinner, a controversial but highly respected Harvard psychologist, wrote a fictional account of a modern utopia. *Walden Two*,[1] Skinner's creation, was quickly praised and damned by critics. Some called it a triumph, others saw it as sinister or appalling. Essentially the critics were reacting to Skinner's idea that all human behavior is externally conditioned. Behavior that is reinforced (rewarded) will continue, while behavior not reinforced phases out of existence. Notwithstanding differences in cortical complexity, this principle operates in man precisely the way it does in rats.

To some, Skinner's ideas on "operant conditioning" have an antihumanistic content. For if man is pushed and tugged by forces external to him, the concept of freedom is nonsensical. Skinner would agree that the idea of freedom, as it is typically conceived of, is meaningless. Humans basically act out impulses to reduce tension and pain and to achieve pleasure. Skinner points to the fact that learning pleasure is a more profound motive in humans than is the threat or experience of pain. "Positive reinforcement" is the only truly effective mode of learning. Or as Skinner's protagonist in *Walden Two* argues:

> The old school made the amazing mistake of supposing that the reverse was true, that by removing a situation a person likes or setting up one he doesn't like—in other words punishing him—

1. New York: The Macmillan Co., 1948.

> it was possible to reduce the probability that he would behave in a given way again. That simply doesn't hold. . . . [Punishment] is temporarily effective, that's the worst of it. That explains several thousand years of bloodshed. . . . Retribution and revenge are the most natural things on earth. But in the long run, the man we strike is no less likely to repeat his act.[2]

Skinner's literary utopian community was based, then, on the belief that punishment is ineffective. Children in the mythological community were not spanked, but they were disciplined by rewarding their correct behavior. The *Walden Two* utopia also encouraged group care of children rather than the traditional idea of parental discipline. In general, the book proposed the "training out" of destructive emotions such as jealousy and possessiveness by "behavioral engineering."

Walden Two represented only a small part of Skinner's life's interests. Long after its publication, he continued rigorous experiments in operant conditioning. Further, Skinner showed no particular interest in founding or participating in a communal society based on his principles.

Yet Professor Skinner's book made a great impact on the campus community, and by 1970, no less than half a dozen groups were attempting to actualize, in one way or another, his principles of community organization. In Grinnell, Iowa, a handful of excollege students began a communal venture based partly on Skinner's ideas. The politically active communalists at Grinnell have an old, two-story house. The second story is reserved for jewelry-making, the production of print washers and other photographic aids, and a photographic gallery. The residents at East Street Gallery in Grinnell give much credit to Skinnerian ideas, but claim no dogmatic approach to their communal attempts. Other communal groups, such as Walden Three in California, also look (as their name indicates) to Skinnerian principles.

The oldest and perhaps most impressive Walden-Two venture is the Twin Oaks community near Louisa, Virginia. It began to take shape as the result of a conference on Walden-Two utopias held in Ann Arbor, Michigan. The participants at the meetings generally agreed that vast amounts of money and planning would be required to actualize Skinner's ideals. This, of course, seemed impossible to most of the would-be-communalists, and they returned to the world of competition, possessiveness, and punishment. Some more adventurous souls were not as easily deterred, however; and soon after, Twin Oaks farm

2. *Ibid.*, p. 260.

began to take shape. By 1970, about two dozen communalists were living and working at Twin Oaks.

The Twin Oaks concept is antithetical to that of hip communes. The inhabitants of the farm look, at first glance, like hip communalists. The males, for example, are prone to wear long hair, and the female members are not given to wearing makeup or the like. But these similarities are superficial, for in many ways the highly structured, anti-mystical Twin Oaks sharply diverges from a typical drop-out hip commune.

The first difference between the two groups is that Twin Oaks communalists do not, like their hip brethren, reject technology. In fact, the Twin Oaks community utilizes all forms of technology that are not dehumanizing or disagreeable. The farm raises pigs, calves, chickens, and a few geese. Most importantly, the community operates a hammock factory, which requires a minimum of technological sophistication and skill. Most Twin Oaks residents work a six-to-seven-hour day and listen to music via stereo or radio while they labor. They seem to enjoy their work. The hammocks are sold for $25, $30, or $35, depending on size. Although they sell rather well with almost no advertising, the profit margin is quite low, and the community is always short on capital. Most money that does come into the commune is invested in improvements, such as a tractor and the enlargement of living space.

The basic difference between the Twin Oaks group and their neighbors in rural Virginia is not, however, the kind of work they perform. It is rather the way in which work is structured.

Krystyna Neuman and Henry Wilhelm, members of the Grinnell commune, visited Twin Oaks in 1969. They reported the following on the Virginia communalists' economic system.

> Central to Twin Oaks is the "labor credit" system of dividing among the members. Briefly described, the community decides each week how much work and what jobs need to be done. Then, people sign up for the jobs they want to do. The number of hours each person must do is determined by dividing the total number of hours of work for that week by the number of people there are to do it. The various jobs are given different "labor credits," depending on their "desirability" or "non-desirability," and this is determined by the number of people who sign up to do any given job. If not enough people sign up to do a certain job, say dishwashing, the labor credit value of it is increased. . . . A person is expected

> to do his share of labor credits—but he may do them at any time he feels like it. You often see some people working in the hammock factory, while others are standing by doing nothing other than enjoyably rapping with the workers, entertaining them while they work.[3]

Men and women share jobs that would be considered manly or womanly in straight society. A visit to the commune may find women in a class in "remedial automobile mechanics," while men cook or do dishes. Twin Oaks, like the Israeli kibbutzim, takes no notice of sex differences in assigning work. In this sense, women's liberation has arrived at Twin Oaks.

Children in the commune, like those in the novel, *Walden Two*, are disciplined by the "child raising manager" rather than by their biological parents. Every adult member of the commune is manager of at least two tasks, since more jobs than people reside on the farm. While it may seem reasonable to have managers for financial problems, some would question the idea of "managing" children. The fact is that the Twin Oaks communalists eventually hope to phase out the biological family as a social institution. Like other groups mentioned in this book, they conceive of the presently constituted family as a regressive social form. It, of course, does foster parent-child and husband-wife exclusiveness, which may become a potential threat to community unity.

The child raising manager, in theory at least, never answers a child's aggression with counter-aggression. Children in the community often find that their disruptive behavior is ignored, not condemned. When, by chance or design, they react in a "correct" way, they are praised or rewarded in other ways.

One of the more colorful managerial positions at the commune is that of the "Generalized Bastard." Kathleen Griebe, a member of the community explains the job requirements for this position:

> His job is to be officially nasty. For example, suppose that a certain member has a habit of letting his work partner do the dirty part of the work and of skipping out on the last ten minutes of cleanup on a shared job. If this happens once or twice, his partner begins to resent it. He hates to say, "Hey, how about doing a full share of this job for a change?" In order to avoid a building-up of

3. Krystyna Neuman and Henry Wilhelm, "A Radical Commune—An Approach to Revolution," *Pterodactyl*, September 18, 1969, p. 9.

resentment, the complaining member goes to the Generalized Bastard. His job is to carry the complaint to the offender, which he can do in an objective way.[4]

The human engineering advocated by Professor Skinner is obviously not completed at Twin Oaks. Human relationships are not always fulfilling. The problem of sexual conflicts or constraints, for example, is not fully solved at this time. One resident of the community told me that some "sleeping around" does occur, but the fact is that Twin Oaks has no formal rules concerning sexual liaisons. Some of the residents at Twin Oaks are single, others are married to their original spouses, and still others "play the field." As Neuman and Wilhelm put it,

> At present, Twin Oaks operates on a more or less "monogamous couple" pattern of sexual relationships. One member said the main reason thus far that people left the community is that they were unable to establish satisfactory sexual relationships. Eventually the sexual pattern of the community may evolve into a more complex pattern which would allow more freedom in sexual patterns. But, again, the strong cultural conditioning of the first generation members may take a long time to overcome. Single visitors of either sex cause problems also, in that they bring with them their unliberated values with regard to the opposite sex.[5]

Interpersonal relationships at the experimental community are guided by a code which states in part,

> We will not discuss the personal affairs of other members, nor speak negatively of other members when they are not present or in the presence of a third party. This rule is both unusual and difficult. Most of us find a certain pleasure in gossiping or grumbling about other people. We feel that this type of talk is harmful to a small community. If a member is unpleasant or lazy or gross, let each other member discover this for himself. . . .

This rule is much like the practice at the Reba Fellowship (the Christian urban commune) of allowing only nonjudgmental comments concerning individuals, during sensitivity sessions. Controlling gossip and

4. Quoted in "Walden Two Lives," *The Modern Utopian*, September–October, 1967, p. 4.
5. Neuman and Wilhelm, *op. cit.*, p. 11.

backbiting is, indeed, difficult, but it is essential in suppressing the disruptive effects of malice.

There are two other important factors in the Twin Oaks strategy of interpersonal relationships. The first concerns the selection of potential members. Individuals who wish to become a part of the community go through a trial or probationary period of three months. If, at the end of that time, both they and the community decide they will fit in with the goals of the group, they become members. In 1970, potential members had always left the community of their own volition. Nevertheless, the group does reserve the right, at the end of the probation, to blackball a potential communalist. This practice would be totally rejected by most members of hip communes, since freedom to drop in or out of any community is part and parcel of their unwritten code. The Twin Oaks communalists are greatly concerned with stability, however; and from this point of view, the trial period is an effective tool for preserving group life.

A second important facet of group life at Twin Oaks is its egalitarianism. Equality is more than just a word at the farm. The educational levels of the communal members varies from graduate degrees to high school drop-outs. Yet no one is allowed to "pull rank." Everyone is called by his or her first name. Moreover, the managerial jobs (especially those requiring decision-making rather than hard labor) are not given particular prestige. The work-credit system also contributes to the lack of hierarchy in the community. Like many other communalists, the Twin Oaks residents see authoritarianism, power, and "ego-tripping" as generally unsatisfying modes of human interaction.

Of course, the problems of the Twin Oaks communalists are not always internal ones. They happen to reside in an area of Virginia infested with the K.K.K., which, characteristically enough, gave George Wallace an overwhelming vote in the 1968 elections. For this reason, Twin Oakers are very cautious about their community's image. They do not discuss the war in Southeast Asia (although they oppose it), or the legalization of pot (although they support this). The use of drugs by the communalists or even by visitors is forbidden. "It's an expedient," explained Kathleen Griebe.

> Adding indignant neighbors to our list of difficulties is more than we can handle. A raid by the police we don't need! Even people off on their private illusions are more than we need when we are trying to get a lot of work done. . . . If we had a population with

> a strong personal desire to smoke pot or drop acid, we might have to take risks. . . . But our membership has virtually no interest in drugs. And if we did have, we wouldn't have the money for it. A liberal stand would benefit only our potential guest population, and we don't see any reason to risk the community's local reputation just for guests.[6]

The restrictions on drugs, together with other efforts toward reconciliation, have produced friendly relations between the rural Virginians and the Twin Oaks communalists. The local sheriff even bought a puppy from the group. This points to the fact that communalists need not always fear the resentment of the local community. Doubtless the communal "wars" in New Mexico owe much to the unstructured nature of hip communes in the area. Many hip communalists would do well to take a lesson in community relations from Twin Oaks operation.

To say that the Twin Oaks farm is structured is not to say that it is authoritarian or rigid. Twin Oaks members have fun in many ways. For example, the "repulsive quartet," a four part harmony group, meets occasionally to sing their hearts out. The group hopes to change its name when the quality of its harmonious soundings improves. Then too, the communalists enjoy talking with each other. For the most part, they are very fond of each other. Some of the two dozen members are, of course, closer to each other than are others, but in the main they enjoy each other's company.

Twin Oaks people are sure they learn from each other as well. One fifteen-year-old girl related that since coming to the community she had learned

> virtually everything I know. Italian history, shorthand, most of my typing, cooking, housework, planting and hoeing, driving a tractor, driving a car, how to use positive reinforcement to handle a small child, and what happens when you try to answer aggression with aggression. Just before I came here I discovered that I really don't have to do much of anything I don't want to—or at least that hardly anybody has any authority to speak of if you want to push them far enough. Basically I can get away with doing whatever I want. But at Twin Oaks I am beginning to believe that in the long run I don't really want to try to get away with everything. Because I've seen other people here act like that and have seen

6. "Walden Two Lives," *op. cit.*, p. 6.

how it affects the whole group, and I think people who act completely selfishly are shits. I don't want to think of myself that way.[7]

While Twin Oaks residents may seem more ambitious than members of other communes, they do not worship work. They work in order to live. Beyond this, they look for the time when work can be reduced to the point where it takes little of their time. When technology increases in its efficiency and productivity, Twin Oaks members will be quite content to expend their time and energy in creative work or interpersonal experimentation. Yet until that time comes, the group will be forced to work for their own survival. In 1970, some members of the commune were forced to drive into Richmond to work on a day-to-day basis[8] to buy needed materials for their hammock-making operation.

The little farm at Twin Oaks has as yet produced no miraculous results, but it has become a model for communal development. It follows basically the communal ideal of small-scale organization and equality. It is true that it is somewhat more bureaucratic than other communal groups, yet it seems there is great latitude for individuals in deciding their roles in the community. Members of the community may not agree with this analysis, but it would seem that much of the commune's ideology is centered around the concept of balance, moderation, or the "golden mean." The Twin Oaks communalists do not reject technology out of hand; neither do they embrace it with open arms. If technology will raise their standard of living without making the communalists slaves of the machine, it is good. Twin Oaks communalists do recognize the need for authority, but they also recognize the possibility of its abuse. Hence, they give managers power, but consistently rotate the tasks each individual is to perform. The Twin Oaks residents recognize the repressive nature of the society around them but do not protest it vigorously so that they will not be harassed. In a word, they compromise when the need arises.

If the community fails, it will probably be for lack of finances or some other external reason.[9] Internally the group seems to have the resources to deal with its problems. The great promise of Twin Oaks lies not so much in its compromises with technology or its relationships with its neighbors. Some of its inhabitants feel that life at Twin Oaks has en-

7. Neuman and Wilhelm, *op. cit.*, p. 11.

8. One member of the commune is a university professor who teaches simply because he likes his work. He contributes financially to the group.

9. In 1970, the draft threatened to take several members of the farm.

hanced their humanness by giving them a more responsive and loving social life. As one communalist reported,

> I have learned to be completely comfortable with, and take for granted, the friendship of over a dozen people. Before I came here I never had more than one or two friends at a time. I don't mean that everyone likes me. But everyone here has a pretty realistic view of me and likes or dislikes me pretty much in proportion to my virtues and faults. There aren't any pretenses to speak of, and very little fear." [10]

10. Neuman and Wilhelm, *op. cit.*, p. 11.

9

HALFWAY MEASURES—SENSITIVITY GROUPS AND COOPERATIVE ECONOMICS

"Love, bumping his head blindly against the obstacles of civilization. . . ."

George Sand

"We can affirm that in the ethical progress of man, mutual support—not mutual struggle—has had the leading part."

Petr Kropotkin

Many Americans are vaguely dissatisfied with their lives. They have little faith in the system of political justice and little sense of belonging to the national community. Dissatisfaction is not enough to motivate an individual to throw away the security of the bourgeois life-style, however. Middle-aged Americans watch communal ventures with a jaundiced eye. They know that a commune may be nothing more than a pleasant interlude in the life of a post-high-school nineteen-year-old. For a forty-year-old with a moderate income, on the other hand, hip, communal life would burn past bridges, and perhaps future bridges as well—with the very real possibility that new communal structures will fail. Typical middle-class, middle-age, middle-income citizens are likely to do a cost analysis of the communal scene and find the cost inordinately high. In a word, the man on the the street is not buying. There are, however, options open to those who are attracted to communal living but who are not ready to sacrifice family or friends to seek the adventure of social experimentation.

Communal life proposes two rather attractive promises to those who would listen. The first is the promise that individuals will be treated as unique human beings in the communal settings. Since communes are small, egalitarian, and antibureaucratic, they can give attention to the individual needs and the creative demands of each person. Another attractive promise of communal living is that cooperation will supersede competition as a mode of social organization. Is it possible to have individual concern and cooperation without full-time communal living? The answer to this question is a resounding "perhaps."

SENSITIVITY TRAINING AND ENCOUNTER GROUPS

By the beginning of 1970, sensitivity and encounter groups were in full bloom in the United States. Small groups of men and women came together in businesses, churches, and student centers. They talked to each other, screamed at each other, and occasionally hugged each other. Almost always, they emoted their feelings toward other group members. They were honest. Not totally so, perhaps, but at least the opportunity for honesty was there. In fact, far more opportunity arose than was often the case when one had to shield his or her authentic feelings to survive in a bureaucratic structure.

Sensitivity training may have been new in 1970, but like most new phenomena, it had its roots in the very old. Mutual criticism, for example, was an integral part of a legion of nineteenth-century communal experiments. The Amana Colonies had their *untersuchen* of mutual confession. Later, more revolutionary groups, such as the Marxist-Leninists, developed "consciousness-raising" sessions. The central question asked in all these divergent organizations was "How do I as an individual relate to the goals and individuals in this group?" Sometimes, if the individual belonged to a highly authoritarian group, he would find the answer to his question in some form of punishment, a harsh rebuff or even expulsion. Other, gentler (perhaps religious) groups might answer the question with encouragement and support. In any case, the question as to how an individual relates to other individuals or to the goals of a bureaucratic institution is seldom asked. Bureaucracies do not run on feelings. In fact, they succeed mainly because bureaucratic workers *put aside* many of their impulses and emotions. As the possibility of working outside a bureaucracy lessened in postindustrial America, the possibilities for emotional fulfillment may have lessened as well. Hence, the runaway popularity of sensitivity groups.

If one were to chart the causes of the revival of the sensitivity or mutual-criticism movement in America, he would probably wish to look into the experiments of Kurt Lewin, a psychologist and refugee from Nazi Germany. Lewin became immensely interested in the way in which groups accomplished tasks given them. In the 1940s, Lewin[1] and his students began to experiment in the new science of group dynamics. Lewin soon found that individuals brought many prejudices, emotional hangups, and feelings of inadequacy into task-oriented groups. Further, Lewin discovered that, as people began to be more frank and open about their problems, they would get on with the task in a far more useful way.

In 1947, Lewin's students, with Dr. Leland Bradford of the National Education Association, founded the National Training Laboratory in Group Development. Training groups (or "T" groups, as they were later called) worked with professional psychologists and counselors to promote an understanding of group processes as well as personal growth for the individual. By the late 1950s, business groups began to take cognizance of the "T" group. The Institute of Industrial Relations at U.C.L.A. began to study the potential of the "T" group for industry. Corporations like Esso began to send their executives to training groups in order to work out their personal problems.

Of course the avant-garde was not long in picking up the "T"-group trend. The Esalen Institute in Big Sur, California, soon developed a nation-wide reputation for its sensitivity or encounter groups,[2] as they were later called. Created solely for the purpose of "enhancing human development," encounter groups at the institute were led by rather unorthodox psychologists such as Dr. William Schutz. The groups led by Schutz and others had no preconceived game-plan; rather, they attempted to follow the feelings of the group, whatever direction that might take. Encounter groups at Esalen usually consisted of a dozen or

1. Lewin's best known work in this area is *Resolving Social Conflicts: Selected Papers on Group Dynamics* (New York: Harper and Row, 1948).

2. An excellent analysis of the sensitivity movement is Jane Howard's *Please Touch: A Guided Tour of the Human Potential Movement* (New York: McGraw-Hill, 1970). Also see Clark E. Moustakas, *Individuality and Encounter: A Brief Journey into Loneliness and Sensitivity Groups* (Cambridge, Mass.: H. A. Doyle Publishing Co., 1968). And, Irving R. Weschler, *Inside A Sensitivity Group* (Los Angeles: University of California Press, 1959).

less people and a trained leader. The leader may remain silent, or he may support, attack, or analyze the group's activities.

Specific tactics for breaking down individual resistance to communication with the group involved a myriad of techniques, some of them zany. At Esalen and other retreats[3] sensitivity sessions were often held in the nude, or perhaps in a large swimming pool. In some instances, participants wrestled, threw garbage at each other, danced, chose new names, milled around, fantasized, or were led around blindfolded. These tactics were no less strange to the initial participants than to the public at large. Some of the tactics were, in fact, created to cause embarrassment to members of the group. Other techniques forced normally reticent individuals to express themselves in emotive ways. Sometimes sensitivity sessions lasted fifteen to twenty-four hours, fatigue helping to break down the inhibitions of group members.

The major significance of the sensitivity movement is that it involves the establishment of a community for lonely, alienated individuals, most of whom are quite normal in a psychological sense. The sense of community sometimes achieved by the group is only temporary. Moreover, many groups fail to accomplish even a sense of emotional support for their members. One veteran of several marathon sensitivity sessions complained to me that "only extroverts profit from the sensitivity experience." Her feeling was that, naturally enough, dominant personalities tended to dominate sensitivity sessions. This problem, of course, is one of leadership. One of the responsibilities of a trained group leader is to draw out feelings and responses from all members, not just those who volunteer their reactions.

Some criticisms of the sensitivity movement are quite serious. Critics have pointed out the possibility for psychotic episodes for psychologically troubled group members. Fortunately, this is a rarity, although it *is* a possibility. Again, this is not so much a criticism of sensitivity sessions as an argument for well-trained group leaders.

Perhaps a more serious criticism of the sensitivity movement is that it has no lasting effects on its participants. Research on this point is scarce at the present. But it does follow logically that attitudes of trust and openness, which are ideally developed during encounter sessions, would not carry over into "the real world" if they are not reinforced.

3. For descriptions of the techniques used at Esalen and other encounter centers, see William C. Schutz, *Joy: Expanding Human Awareness* (New York: Grove Press, 1969). Also see Rasa Gustaitis, *Turning On* (New York: Signet Books, 1970).

And attitudes of trust, openness, and love are not reinforced in the business world, where honesty (in a total sense) is not the best policy.

In essence, the sensitivity-encounter sessions can be seen as a Dionysian orgy *without sex*. This, naturally, would fault bourgeois sensitivity sessions, for most hip communalists see sex as a completely natural way of communicating. More than this, hippie love tribalists trust the touch of sex far more than the sometimes devious verbal webs woven by argument and discussion. Hence, when two girl members of a hip commune in the midwest leave the room and ask anyone who wants to "get laid" to follow them, they are expressing (besides sexual desire) their willingness to communicate physically with any member of the tribe. Love in the hip commune is often liberated from permanent commitment. Certainly, it is liberated from emotional "cost analysis."

The costs for belonging to hip communes are well known. Besides the loss of a comfortable, middle-American existence, it could involve the contraction of venereal disease, such as was the case for several members of Kerista, a free-love communal group in California. Other less than pleasant possibilities are bronchial diseases or the less serious, but more dramatic, C.C.'s (communal cramps) due to the dietary problems of communal living.

Thus, the hip communalists berate sensitivity group members for copping out by not dropping out of their comfortable, sanitary lives. Most sensitivity members look at the costs paid by hip communalists and reply "too much."

Robert Claiborne, writing in *The Nation*,[4] is essentially critical of the "human potential movement." Much of the movement is essentially a new opiate of the middle classes, Claiborne argues, "a new toy pacifier for affluence."

Yet, the movement does have potential according to Claiborne.

> The liberating possibilities of human potential groups are suggested by a number of things. First, there are the unfortunate experiences that some corporations have had with it. A *Wall Street Journal* survey cites the cases of several businessmen who . . . "went into "T" groups autocratic, tough, and competitive, and came back so radically changed that they lost their jobs." According to the *Journal*, many corporations have, therefore, "turned away from the free swinging, traditional "T" groups to carefully

4. October 19, 1970, pp. 373–74.

> planned sessions dealing . . . with specific company problems." And a General Electric spokesman is quoted as saying that sensitivity training must be designed to change an individual, not necessarily change the environment he works in—to whittle the peg, not redrill the hole. A movement that can, even occasionally, radically change autocratic, tough businessmen can't be written off as merely a toy; what's bad for General Electric may be good for the rest of the country.

Claiborne argues further that sensitivity training must confront the social system if it is to have value. Here, his argument runs much like the criticism radical political communalists make of their hip counterparts. Simply "grooving" on other people is no guarantee one will avoid oppression. Thus, Claiborne sees the valuable group experience as radicalizing.

> A prototype of what a radicalizing human-potential movement might be like can be found in the "consciousness-raising" activities of some Women's Liberation groups. These meetings . . . are devoted to discussing the participants' experiences as women, and in particular the personal frustrations and injustices they have suffered on account of their sex. With admirable concreteness, they concentrate on the particulars of their own lives, as a means of understanding the problems of women in general. A still earlier analogue can be found in the device invented by the Chinese Communists (now, of course, swept away by the cult of Chairman Mao)—the "speak bitterness" meeting. Here, peasants were encouraged to recount publicly their own mistreatment at the hands of landlords and bureaucrats, as a way of strengthening the entire village's determination to resist oppression.
>
> In both cases, clearly, the aim is much the same as in many human potential groups: to release people's real feelings about themselves and their lives. But the more politicalized groups do much more, for they seek to focus the released feelings of anger, frustration, and bitterness on the specific individuals and institutions that have engendered them, thereby fusing personal consciousness with radical, revolutionary consciousness.[5]

5. *Ibid.*

ECONOMIC COOPERATION

If the communal ideal of emotional support can be realized at least partially by sensitivity sessions, it is also possible that the communal sharing of material goods can be accomplished partially by producer and consumer cooperatives. The idea of a cooperative rather than competitive economic system goes back to man's dim past. Certainly by medieval times, small societies such as that at Tabor in fifteenth-century Bohemia had begun to organize themselves on the basis of economic cooperation. Their attempts, like those of the medieval Albigensians, were primitive. They sought pure communism, the sharing of all with all. This can be done today, and pure communism does exist in a few hip communes. But pure communism necessarily limits the standard of living so drastically that even few of the hip communes are "pure" in their economic sharing.

An alternative that lies somewhere between the competitive, almost cannibalistic, corporate economy of the United States[6] and the drastically egalitarian demands of pure communism is the cooperative movement.

Cooperatives are not new in the United States; many were formed by the populist movements in the latter half of the nineteenth century. What is new about current cooperatives is that many young, long haired, counter-culturists are experimenting with them. Young artists, who may or may not live in a communal setting, often pool their arts and crafts for cooperative selling. Even more popular is the idea of cooperative buying. The *Eugene* (Oregon) *Register Guard*[7] describes a hip co-op in Willamette, Oregon.

> Dan's wife makes the bagels; Mrs. Neusihin makes the pickles. Granola, "a cereal that really stays crisp," is 60 cents a pound packaged; if you bring your own container it's 55 cents, and you can buy as little as you need.

6. The ideology of early capitalism was that the economic struggle for existence benefited man's evolutionary progress via natural selection. Prince Petr Kropotkin, a Russian aristocrat and zoologist, challenged this idea with his book *Mutual Aid* in 1902 (republished in 1955 by Extending Horizons Books, Boston). Kropotkin did not deny the validity of the natural selection principle, but he did maintain that evolutionary development and progress owe more to mutual support and cooperation than a refusal to care for the weak.

7. February 25, 1970.

Willamette People's Cooperative, the corner grocery store which sells these items, is a booming business. Two months old, the grocery at 22nd and Emerald has over 500 members (at $5 a share) and is grossing $700 to $900 a day. Already there is talk about starting another co-op to handle a volume which surprises even the organizers.

The place was started by a group of University of Oregon students and their friends who wanted to sell groceries at lower prices and to sell an idea—that a sense of community can be created through a common cause and need. Transportation problems have blighted that community spirit somewhat, but volunteer sales help (20 to 30 clerks who work without pay) continues strong behind the counter.

The co-op buys a lot of its stock in Portland and from farmers, slaughterhouses and wholesalers around the country. People have given cars (some of which won't run) to the store for pick-up runs by volunteer drivers. But occasionally you won't find the Tillamook cheddar which sells at 81 cents a pound or fresh eggs for 66 cents a dozen. . . .

After two months of successful operation, the co-op membership has voted to shift the emphasis to more natural and health food. At a meeting last week members decided to stop the sale of cigarettes because smoking is "a filthy, addicting habit"; to stop the sale of "garbage" sweets and stock healthful candies for the school kids; to limit stocks of certain packaged foods which aren't particularly healthful; and to emphasize the sale of fresh fruits and vegetables over canned goods.

Jack Corbett, who works afternoons at the co-op, says rumors that the store is being harassed by food licensing agencies are untrue. He said the clerks who cut the cheese and meats for customers have to have food handlers licenses, and Corbett agrees they should. He said he's found the food inspectors helpful, and it's simple to meet health requirements if you're willing to listen to inspectors. . . .

Shoppers include everyone from youngsters who pay seven cents to help themselves to one of Mrs. Neusihin's pickles to old timers who like the atmosphere. Browsers read signs explaining the benefits of natural grains and poly-unsaturate oils, the average wage of Guatemalan farm workers, and the number of war victims in Vietnam. Buyers help themselves to flour from wooden barrels and grind their own coffee. The adventurous can even take home a piece of horsemeat to try out on the family.

The hip co-op is only a partial rejection of capitalism; it does not require the rigorous material asceticism of the hip commune. It is possible that co-ops such as the one in Willamette, Oregon, do give counterculturists a sense of control over their economic life. Perhaps more important is the fact that cooperatives require planning and social organization. The "turned on" young men and women who planned and organized the pioneer co-ops in Berkeley or Madison have confidence that they can create a *successful* set of alternative institutions to challenge those of the straight society. To an extent, they have already done so with the underground press. If the cooperative movement continues to grow, as it is likely to, it could produce serious competition for local businesses.

Of course not all of the new cooperatives are organized by "longhairs." Five housewives in a hard core poverty area of Brooklyn, disgusted at the high prices and poor quality of food, organized the Cuyler-Warren Consumer Buying Club. The Buying Club was praised by Mayor Lindsay and proved in its initial stages to be a huge success. Also the School of Living, an established communal research group, created the Devcor Farmer-Consumer Co-op, in 1970. Devcor is basically a buying club, which means that individuals, with the aid of the organization, buy directly from the producer. In 1970, Devcor offered fruits such as apples, oranges, and peaches, as well as vegetables and eggs.

The cooperative movement is a natural direction for the young counterculturists, since it rejects the bureaucratic middleman, emphasizes the cooperative rather than the competitive, and allows individualism in the market place. Moreover, capitalism has been severely criticized for its incessant creation of new demands and artificial needs within the individual. Literally thousands of advertisements bombard the consumer with the idea that he should not be satisfied with what he has or what he now desires. This relentless propaganda has its effects—Americans buy; and much of what they buy is junk. Its advocates say that the promise of the cooperative movement is that it does not depend on the sometimes dishonest, often hyperbolic, system of advertising.

The cooperative movement is undoubtedly more acceptable to members of middle America than the radical life-style of hip communes.

Tolerance for long-haired youth has its limits anywhere in the United States, however; and some "freak" cooperatives have been the target of vandalism. The owners of the Bowery General store in Iowa City, Iowa, sell organic foods to a chiefly hip clientele around the University of Iowa. They live in a rural area east of the city. On October 14, 1970, the owners of the store, Luther and Anna Danneman, were in Wisconsin buying a stock of honey for the store. A long-haired friend of the Dannemans, Michael McKaie, slept in his truck on their property. The *Des Moines Register*[8] reported McKaie's experience.

> Something woke me up. . . . I got up and looked around. Cars were parked all around the place, and men were milling around, some in the yard, some out on the road, some in the house, and I heard things being turned over. I got out of the truck to talk to these men, and I asked, "What's going on?" There were 15 or 20 of them, and they started getting rough right away. My hair was tied in a ponytail, and one guy grabbed the hair and started reeling me around. Then they stood around in an arc, and they started talking to each other. "Look at his hands; they're dirty," one of them said, and, of course, my hands were dirty. I'd been working on my truck, and another said, "He's the kind of guy who has no respect for the flag."

The mob proceeded to hack at McKaie's long hair with a pocket knife.

> After that, one guy threw me into the side of the truck. It wasn't a painful blow, but it stunned me, and I fell down, and someone said, "Get up, you're not hurt," which I wasn't. I reached up to my head, and it felt real damp, and then blood came down my face and onto my jacket, and the men seemed to get uneasy, and all of a sudden they split.[9]

McKaie believed the true target of the raid to be the owners of the natural food store. The local paper berated the attackers, and eighteen students from West Branch (the small town closest to the incident) wrote a letter of apology to McKaie for the attack upon him by "the so called respectable men of our community." Not all of the community was repulsed by the vigilante action, however. One neighbor of the Danneman's verbally attacked the longhairs rather than the vandals.

8. November 22, 1970.
9. *Ibid.*

> I feel just like Mr. Nixon, the president of our great nation, and someday it will read "once was a great nation." . . . Mr. Nixon says the people that do the heckling and are hollering the loudest will not be the leaders of our country, God forbid. It is people like these hippies that are going to bear freak children because of their dope when they were younger. Would you tell me who is going to pay the care of these people . . . ?

His wife added,

> That poor hippie that got in trouble didn't belong there either. The renters were not there. I suppose, by rights, we don't have the right to say we don't want these people in our neighborhood; is that right? [10]

Hip entrepreneurs and co-op managers in rural and urban areas expect a good deal of dislike. Surprisingly, their greatest dislike has not come from commercial groups in potential competition with them, but from those who dislike the *symbols* of the counterculture—long hair, beads, buckskin jackets, and peace symbols. Hip co-ops can survive without the approval of intolerant neighbors, if they are not "busted" by the legal authorities. Hip co-ops, like hip communes, are growing, and unlike many hip communes, the co-ops are well-organized and thriving.

10. *Ibid.*

10

THREATS TO COMMUNAL LIFE

"There comes a time in the Affairs of Man when he must take the bull by the tail and look the situation squarely in the face."

☺ *W. C. Fields*

"All the stones are cut to build the structure of freedom; you can build a palace or a tomb of the same stones."

☺ *Saint-Just*

A betting man, at this point in history, would probably be willing to make good odds on the failure, within the next ten years, of most communes now extant in the United States. It could be pointed out to the odds-makers that more young people are currently dropping into communes than are dropping out of the communal scene. It could also be pointed out that some hip communes are better organized than before, and are developing a new sense of dedication to group living. Still, most dispassionate observers of the communal scene would probably agree that the "smart money" is against the continued existence of most communes, hip, religious, or political. Communes are threatened both externally and internally; externally because communalists have little power in their dealings with the outside world and internally because of the possibilities of decay from within.

EXTERNAL THREATS

In 1970, most people in the United States were either uninformed about communes or unconcerned about their existence. With problems such as unemployment, war, pollution, and violence on their minds, most Americans could care less about the communal movement. It is true that many of the symbols associated with communal living, such as dirtiness, drugs, and free sex, were, and still are, antithetical to the beliefs of the typical American housewife in Dayton.[1]

In areas with heavy concentrations of communalists, such as New Mexico or northern California, the anger of local residents smolders. "Why are these people allowed to stay in our neighborhood? What about property values? What if our kids begin to mimic those hippies?" These complaints, reminiscent of the feelings of suburban whites about the invasion of black people, are real; and if the fears of local residents are not abated, they can grow into outrage. Local law officials often share this distaste for communalists, who go against the grain of middle-American values. Moreover, local law enforcement personnel are, of necessity, sensitive to the mood of their community. Dispensing justice (to say nothing of mercy) to an unpopular group may well threaten their jobs. Further, it is an unpleasant fact that dissenting groups nearly always received a less than tolerant reaction in early America. Thus, it is no surprise to learn of incidents such as the one at Oz, a commune in rural Pennsylvania. A few months before the hip commune disbanded, it was invaded by a group of "bikers." The nonhip populace is likely to regard bikers or motorcycle gangs as simply mobile hippies. They are not. Rejecting the love ethic as unmanly, bikers in many parts of the country have terrorized both straight and hip communities equally.

The gang that attacked Oz was no exception. Showing equal amounts of sadism and lust, the cyclists proceeded to beat and torture the male residents of Oz, as the females were slapped around and sexually assaulted. After dark, the terrorist gang proceeded to force the male communalists to walk through the still-burning fire for entertainment. Girls hiding on the second floor in the dark were systematically sought out by candlelight as prospects for rape.

The communalists had decided that they would not (and, in all probability, could not) resist with force. Beyond that they were undecided. One male communalist "freaked out" in hysteria. Finally, a member of the commune reached a telephone and called the local police. The police had never pretended to admire the pioneering spirit of the communalists, with its emphasis on freedom, drugs, nudity and the like. It need hardly be mentioned that they took a dim view of

1. In 1970, the Gallop poll discovered the average voting American—a female, 47 years old, married to a machinist and living in Dayton, Ohio.

bikers as well. Hence when they arrived, they did not exert themselves to a great extent on the communalists' behalf. That is not to say that the bikers escaped with no punishment. Before they left, the police forced them to pick up many of the beer bottles they had strewn over the communal area. The police, it seemed, were not fond of litter-bugs.

Incidents such as this have happened to communalists in many parts of the United States, and they remind one of the lenient policy many Southern and some Northern police have had toward the ghetto populace. "If one black does it to another one, well what can you expect?" A number of studies have shown that black crimes against blacks are less severely punished than black crimes against whites, or for that matter, white crimes against whites. Why is this true? Basically because black people represent a minority group—that is, a group with less than proportionate political power. Hip communalists are a minority group in this sense too, except that their minority status is a voluntary one. For this reason, hip communalists are not likely to get a fair shake when they confront middle class judges, lawyers, and police. A certain amount of power is a necessary requisite to equal justice. For this reason, local hip communalists and street people have, on occasion, become political. In Berkeley, in 1970, they fought for a referendum on community control of police; and in Aspen, Colorado, the November, 1970, elections saw hip politicians running for sheriff and justice of the peace.

There is a paradox in all this, and it is this: to be left alone in this or any other country, one must seek a certain degree of political influence. Seeking influence, of course, lessens the possibilities of being left alone. Yet there seems to be no way out of this dilemma. In some instances, communalists may live in highly tolerant areas; they may not flaunt their differences with the larger population. These cases are probably unlikely, and in any event, communes need the protection of the Bill of Rights of the United States Constitution. It has been said that a majority of the American people would reject the Bill of Rights if it were put to a referendum. If this is true, it points up the need for some kind of political consciousness among communalists. If repression does come, either from local citizens or authorities, only effective political action can save the communal venture. The chanting of mantras and "Oms" may produce spiritual purification, which may be little protection against eviction notices or bricks thrown through windows.

As well as legal and nonlegal persecution, communalists, if they are serious about group survival, must keep lines of communication open with the outside world. They can not be ignorant, for example, of the recent technological achievements in farming. Finding pure water, growing enough to survive, and living off the land take skills, and these skills can not be acquired by ignoring the larger society. Communalists do not have to emulate the waste, greed, and unhappiness so evident in the marketplace, but they should not regard all technology as evil. If they were to do so, they would probably find their dreams of freedom ossified with the rigidity of the horse-and-buggy Amish.

Because of their nonrevolutionary goals hip communalists have not been the target of Federal scrutiny or prosecution. Such is not the case, of course, with the revolutionary urban collectives. In 1970, four of the top ten of the F.B.I.'s "most wanted" list were members of revolutionary collectives. Two of the four were young women. All had been implicated in terroristic activities. They, like many of their confreres, considered themselves urban guerillas. As the government developed a greater interest in the political crimes of radical young people, there was great fear in many quarters for the vulnerability of societal institutions. Bombings and abortive kidnaping plans gave the law-and-order forces near *carte blanche* in hunting down the young organizers.

It is difficult to predict what will happen to the political collectives in the future. Yet, if terroristic tactics increase, or even the threat of those tactics increase, repression will certainly follow. Further, activities by collectivists such as the Weathermen have, in the past, been elitist, since they had almost no popular support.[2] If the future replicates past action in the United States, both terrorism and repression will escalate, just as every massive troop build-up by the United States in Vietnam brought counterescalation by the North Vietnamese. Violence is a dialectic; it produces its own antithesis. It would seem that revolutionary urban communalists and civil authorities are increasing their readiness to "play the violence game."[3]

2. Political revolutionaries are often condemned by hip communalists as well. Gloria R., a sometime communalist and Women's Liberation supporter, wrote in 1970—"I was bothered by the ego-tripping evident within the movement, especially among the people who advocate violence. Violence in this society is an ego-building, masculine trait, while love and gentleness is the sign of a 'sissy.' I'm afraid that even the revolutionaries haven't been able to shake this basic indoctrination. The total revolution is going to be a long time coming. If we force it on the average American, we're no better than the Establishment!" *New Prairie Primer*, November 23, 1970, p. 2.

3. A tragic account of a young revolutionary's entanglement in violence and eventual death is given in J. Kirk Sale's "Ted Gold: Education for Violence," *The Nation*, April 13, 1970, pp. 423–29. Gold, a young idealist, who joined a Weather-

Religious communalists have least to fear from the Federal authorities and from their neighbors as well. Nevertheless, most religious communalists described in this book have political concerns. Religious communes, such as those founded by the Quakers, Catholic Workers, and Mennonites, have a long tradition of nonviolence and official pacifism.

In summation, modern communalists will be harassed and suppressed to the degree they are perceived as a threat to the larger social order. Some of the perceived threats imputed to communes are silly. Question: "What if everyone dropped out and joined a commune?" Answer: (equally silly) "Well, first the war in Southeast Asia would end immediately, the arms race would stop, destruction of our air, water, and land would end, etc." The fact is that a majority of Americans will never participate in communal experiments. Further, with the continual crisis of overproduction in American society, the nonproducing communalists represent no threat to the basic economic system.

Many parents are deeply concerned that their own children may join a hip commune and forsake their birthright of affluence. This is a legitimate fear, of course, but it must be tempered by the fact that one generation can never imprison the next with its particular goals. Change is far too ingrained into American society for that. Some parents who have sons and daughters living in communes are able to accept and love them regardless of their differences in life-style. This is difficult for protective parents to accept, but doubtless more satisfying than cutting one's self off emotionally from one's own children.

For their part, communalists must walk a delicate line by attempting not to offend their neighbors unnecessarily, while at the same time preserving their sense of uniqueness and integrity. Occasionally this may entail compromise. At other times, hip communalists must make a stand for their principles. Hip communalists who are unflexible in their life-style would do well to pay heed to Lao-Tsu who said some 2,500 years ago that softness is life and hardness (resistance to change) is death.

INTERNAL THREATS

In a mimeographed brochure, members of Greenfeel, a free love commune in Vermont, expressed their hopes and beliefs.

---

man collective, was killed in an explosion in a Greenwich Village townhouse. Officials claimed Gold and his fellow Weathermen had over 100 sticks of dynamite, some packed into twelve-inch lead-pipe bombs.

> It is time. We are a group of people who want to live in a new way together. Each of us has fought a hard personal war—against parents, the school system, society—against fear, self-destruction, loneliness. We've won several battles in the Good fight, and we want to change ourselves more. But we no longer need to fight alone. We want very much to join each other as a community of lovers. Beautiful, alive human beings who work, play, create, experience, and love together. . . . We want to regain the animal use of our senses, making love to life. . . . To love yourself you must love your body. To love your body you must touch and be touched. Sex is yours, and sex is clean. Couples walk by who look secure, but I see how distant they really are. You can't walk up to a couple and say, "I want to be close to you; let's love each other." It's asking them to break their psychological bonds. I can't believe that's love. It's not real. Such limited love. Such fear of change. It is a lie to have just one lover. You have to pretend you don't desire all the other beautiful people. Possession is like a pair of boots which are glued to your feet, and you never, never get to touch the wet grass.

Despite its effulgent rhetoric, Greenfeel is now dead. One of its former members wrote me that "We are now scattering to meditate about the happenings at Greenfeel. Right now we're on R and R [rest and recreation]."

Why, despite its seeming openness and love, did the group fail? No one can say for sure, but the etiology of communal failure usually revolves around one of four issues: lack of leadership, lack of means for handling internal disputes, lack of ideology, and external repression.

It is true that communes are generally egalitarian in all their accouterments, but this does not preclude the functioning of natural leadership. Nearly every hip commune has its charismatic leader, no matter how disorganized it may seem. Hip communalists have looked up to figures such as Ken Kesey, Ramon Sender, Lou Gottleib, and Bill Wheeler. Leaders such as these have been able to mediate quarrels, fight legal interdiction, and generally provide a sense of authority and trust for their fellow communalists.

Michael Olmsted's book *The Small Group*[4] proposes that one function of a leader is to see that loves and hates do not get out of hand,

4. New York: Random House, 1959, p. 141.

> . . . There is the problem of controlling the resentments and hostilities that are likely to exist both by virtue of personality conflicts and by virtue of the friction generated in the course of arriving at group decisions. Among the integrating processes useful for dealing with these difficulties are repression and projection. . . . [This is accomplished] by a scapegoat or by someone who—by jokes, righteous indignation, or other means turns the hostile impulses outward to external objects.

Another attribute of the good leader is that he is

> ". . . the good example, the individual who provides a model of an unconflicted (that is, well-repressed) personality." [5]

Many times potentially explosive conflicts in communal settings are mediated by personalities such as those described in the above paragraph. Lewis Yablonsky cites the case of George Gridley, a natural leader at several communes in California. Near Malibu, one communalist spoke bitterly about the commune's problems and his desire to "cut out." Gridley spoke out vehemently.

> You fucking people are disgusting. Here you have this beautiful mountain, you have no hassle with the country, no hassle with people trying to bust you—you got it made. . . . Why can't you get together? . . . Don't quit, man. Make your light shine. Here is where it's happening. It's happening all over. . . .[6]

A discussion of the possibility of giving free food away followed. Some maintained that someone should stand guard at the door to line up the hungry free "customers" to guard against some getting all the food.

> Gridley disagreed: "No, man. People should learn to control themselves."
>
> At one point in the discussion, Gridley made a demonstration of one of his viewpoints. Someone asked him, "Can I have one of your cigarets?" Gridley fumed, "Fuck you, man. Don't you know that these cigarets are yours as well as mine? When are you going to learn that brothers share everything." The man said, "Too much.

5. *Ibid.*
6. Lewis Yablonsky, *The Hippie Trip* (New York: Pegasus, 1968), pp. 82–83.

You're right. That's right. Your cigarets are mine," and he took a cigaret and lit up.[7]

Hip communes, like any other form of social organization, require leaders who can give stability, hope, and trust to the group. Individuals who have dropped out of straight society may have a tendency to drop out of hip communities when the going gets rough, as it inevitably does. The leaderless commune is doomed from the beginning. The real threat to the comparatively stable commune is that its natural leaders will tire of their group responsibilities as moral exemplars. Decision-making about the economy of the group requires leadership as well, and if the communalists wish to become economically independent from the capitalistic system, crucial decisions must be made. If no one facilitates the decision-making, the group may dissolve.

Generally, hip communalists and other *avant-garde* utopists are threatened in their communal development by two factors—drugs and sex. Both, as any communalist will tell you, are fun. Nevertheless, both are potentially disruptive to community life. Thus, Griscom Morgan has reported,

> As we add up our experience with groups that have extensively used drugs, a number of things stand out: with the release from tension, up-tight social controls and standards, there is some increase in spontaneity but also a general pattern of demoralization, basic economic incompetence, and failure to keep house and to maintain essential sanitary conditions. Behind all this lies the common observation that under drugs individuals tend to increasingly "turn off" their relationship to society, including family, community or commune and its members, and to be turned inward to their own psychic life at the expense of effective mastery of outer relationships. The long term results in commune life are degeneration of social life, a short life expectancy of the commune, deteriorating mental and physical health, and general demoralization.[8]

The use of psychedelic drugs does not inevitably lead to personal and social disorganization. The plains Indians of the United States have long used peyote, an hallucinogenic cactus, in their religious ritual. Their use of peyote has not been deleterious to them, according to

7. *Ibid.*, p. 86.
8. Griscom Morgan *Intentional Community Handbook* (Yellow Springs, Ohio: Community Services Inc.), p. 86.

most anthropologists. It is clear that the use of nonnarcotic drugs and alcohol are not harmful to individuals *if they are used in conjunction with religious rites.* This, of course, is not the case with most hip communalists. Many longhairs and freaks enjoy experimentation with drugs in the same way they enjoy experimentation with sex. They are not consciously trying to reach some new transcendental consciousness, but merely the desiring new experiences. Although drugs heighten perception of color and appreciation of music, social perception is likely to be detrimentally affected. For example, many "stoned" individuals begin to perceive their ideas with an unreal sense of profundity and wisdom. Also, hip communalists do not often perceive the need to regulate drug use so that it does not interfere with the communal workday. For those communes trying to eke out a living by subsistence farming, drugs are often a very real source of trouble. A number of hip communalists have posted signs reading "No drug use here!" This is done partially because of fear of law enforcement authorities and partially because excessive drug use means slow death to communes. Consider, for example, the case of Kathy A., a student at a large midwestern university. She spent two years in an odyssey of communal living across America. Having now returned to the straight world, Kathy attributes many of her problems during her communal living to the constant and excessive use of drugs. I asked Kathy to describe her years in communal living, as well as her drug experiences. The following passages, taken in large part from her diary, recount her drug experiences as well as her increasing disenchantment with hip communal living.

A COMMUNAL ODYSSEY

by Kathy A.

"I pulled an exceptionally large potato chip from the pack I'd just bought. Somewhere in a potato chip factory there was a stoned freak who'd made that and now here I was, a stoned freak, finding it. Another union was complete. Later I was in a garden court at the university. Toby was sitting beside me, but Toby was also in France. He had come into the court, and, as though he didn't know me, he'd asked to sit with me. I couldn't look at him for fear recognition would mean he couldn't be in both places at once or something. There could be no other reason for his refusal to say that he was Toby. In the evening Marco and Parmel called me; they wanted me to come back to Pennsylvania. I told them I had to go to school again so that I'd be able to support us all. I'd been there a week and hadn't even seen the admissions officer. I couldn't because men had been equipping the house next door to watch me through the TV set. They were trying to destroy the chances of us ever starting another commune in the United States."

Actually this was the end of communes for me. The beginning was in California two years before. Those two years at Ben Lomand, San Francisco, Lucerne, Pennsylvania and New Mexico communes brought me to the mysticism and paranoia of the preceding paragraph. It's those feelings I wish to explain—the mental and emotional tendencies I, and I think others, experienced in this unique living situation.

Here is an entry from my diary the day we found Holiday near Ben Lomand, California, and decided to stay: "So we came here in the Santa Cruz mountains. There were fourteen walking the road, holding each other, carrying axes and a jug of wine, men and women both, hitch-hiking. They took us home. The road bends down brown between a sun field and river valley. A thin man with blond beard is hoeing in the garden. They jump out of our back seat, the trunk, off the roof, the hood and embrace us. We are living with them and others, travelers international and local, three week festival guests, and one with a round fuzzy face, Tracy, who said his horoscope indicated he was on another planet. My brothers and sisters are too much. There are six cabins, all knotty pine with swinging double doors over that river with three waterfalls and rock banks. In the redwood forest a clearing is made for meditation. Each cabin is our home, too; like going to another room in the house you grew up in except now the house covers 6½ acres and looks like an old resort. And now everybody in your family hugs each other in a huge swaying mass of 40 energies ohming and cooks dinner together and rides around the hills and to the beach in a bus."

It was here that I experienced the first council meetings where everyone had a voice but few decisions were made, where dope was shared and speed was generally despised but never outlawed for fear of obstructing freedom and where everyone dug through grocery garbage cans while one or two tended the garden or cooked the dinner or built the fire or washed the dishes.

A few months later I was inclined to write, after a celebration:

"It was at this point then that the castinets were calling And each invented a drum to dance the dance of his name. Eyes into eyes with smiles Bright bodies reflecting the fire. At dawn the dove in the redwoods was no surprise. But the amazon And the astrologer Just in from eternity Their mirror faces Their instant knowledge of our wildest dream Some found uncalled for And left us separate again."

It was a feeling of loss of identity needed and not found. I think we all wanted to feel we were a whole body together, inseparable. We tried to prove it by moving en masse to Pennsylvania where the gift of a farm awaited us.

Diary entries there: "Tonight we yelled 'ice cream—you scream' off the front porch to the citizens' motorcade, and three gallons were manifested. Basically the people are lovely (this was before I was met at the door of a small grocery in the next town by the grocer and his butcher knife which meant I couldn't come in), some bringing food and beer, others just stand around and watch us. Another day of TV men—a spade cat and beerbelly from Pittsburg who filmed Marco playing his dulcimer and talked with Mike, Rebecca, Amy-child, Dan and Hickory. The doctor came too and said if we were growing any weed he'd have us in the 'pokie' in an hour. Still cooking outdoors over an open fire—sometimes in the rain—bannock and rice. Put in the rest of the cabbage today with two fishes—will see what the Indians have to say. Also planted parsley, oregano, thyme, mint, lavender and caraway. Then pansies and petunias. The Cleveland Plain Dealer came this morning and took a family portrait—will send us pictures. Many presents today —ice cream again and much greens from the produce man in Meadville, pineapple, grapes, ryekrisp, peanut butter, bananas. Food trip! Still no dope since we got here. Muffin said today that we astral trip in dreams—she has had the same dreams as David before and met some cats who had all met in their dreams before they came to know each other in the physical plane. Last night I dreamed I met some rabbis—2—in a sunken church—they had long white beards though one was younger than the other, and we waved our hands over a watch and made the hours fly. Nice spades here this morning who turned us on to free food supply in Meadville. Started crocheting a thing tonight—not sure what it is. Last night Debbie and Weird Phil saw a poltergeist in their room in the form of a candle-size flame suspended in the air. Local drunks brought us 4 gallons of Muscatel and tons of beer. We all got quite high while they got drunk and had an incredible jam—Muffin sang a while and played guitar—'Just keep my grave clean' and 'It ain't necessarily so.' Then Mike on recorder, Marco's dulcimer and balding John with a sweet licorice stick. Pioneer Virginian and her brood from the farm brought us clothes, candles and walnuts and reminded us the most important thing is to keep our minds clean. Joe brought Adolph and Deja home with six loaves of bread she had made and mucho chocolate chip cookies. But the highlight of the evening was Bill Nordquist—deputy sheriff of Crawford County who came on his own 'as a person' to find out about us and himself. He figures we're doing what everyone wants to do but doesn't have enough guts to do—himself included. David rapped to him about the universe and how God makes everything happen. But old Bill knows where it's at—he carries no gun on or off duty and hasn't made an arrest in two years. He'd rather have people understand the law than bust them. Before he left he said he'd felt he'd come home. Late in the night more drunks—one passing out in our living room, then animalized, growling and whining down the road. Early morning Anthropodius and Josh come home! This morning wanted to get the Nana-Banana nanny goat at the Meadville auction, but it looked like rain and no ride. Last night the town meeting with lots of yelling but at home a high on Ohm. Muffin and I went for mail with far-out Fay in her rum-rum. George was on the radio at five with sympathy for the townspeople's fear of us, yet tonight Ross and three others were thrown out of Conneaut Lake Park by four state police, in the county with the highest suicide rate and third highest crime. Yet this morning the state dick was right on top of our phone call about Chico shot through the paw by who knows and said good things in our favor. Patty-Po and Deja and Adolph talked to the ouija who said we would move and then threw the tarot for us."

And the tarot cards said we would move too. Which we did—right into jail. We were arrested without a suitable warrant and subsequently charged with maintaining a disorderly house and contributing to the delinquency of minors. At a hearing we were persuaded to accept an agreement that we could not live in the county for the next year and that charges could be reopened any time during the next two years. After we left, the house was burned.

Approximately 50 of us divided into small groups and went

different places. We tried to keep in touch. I lived in Iowa a while, then the East-West House in San Francisco, with friends in Santa Cruz and then spent quite a bit of time at a one house commune in Lucerne. There things were even more communal than I'd experienced before. All clothes were put in one place for everyone and mattresses formed a big circle in one room for sleeping. It was a good deal more organized than Oz or Holiday, but then there were fewer people and more communication.

Diary there: "Spirit is contagious. Living so close to one another our thoughts can intermingle or collide. How many thoughts are my own, origin my soul and how many from others? Listening to the multiple phases of mind. How to use the red light? How to know is not to know. Better they say to accept—the hate, the jealousy, the desire. Better to flow with arms open, flying into the situations which will only return again if rejected now, return with the added sting of a past mistake. The red light identifies, accepts, and overpowers. And I am free again—to love, to consummate, to make my spirit contagious with joy. This morning Patti raps about the continuum of our past lives; many manifestations of the soul coexisting in the everpresent now, many beings at once. (You are me as I am he and we are all together.) But this illusion of separateness. I meditate this morning in a room with three other people and must work very hard to force that illusion of me up my backbone. And yet felt very comfortable, unhindered yet requiring more concentration than normally. Last night too many Leos for me to handle—had to exit, go to bed. Then into that trip of guilt for not facing the difficult, not flowing."

I continued to feel guilty. All the mysticism of our occult practices meant to us that we should live in harmony with one another. All the drugs we took, at least for me, set us further apart. Eventually this town rose and threw us out, too. The house was condemned. We decided to move to New Mexico, but we were hardly welcome there even by our own people. It was there that I felt the effects of my guilt. I was beginning to doubt that it was possible to live closely and harmoniously. I was constantly stoned, and I was undernourished. My dreams had died. I remember that I did a lot of reading those two years, much of it concerned communes. I read a history of Brook Farm, one of the first communes in the United States. Many prominent people were there trying, as I think we hoped to try, too, as George Riply, the founder said, "prepare a society of liberal, intelligent and cultivated persons, whose relations with each other would permit a more simple and wholesome life than can be led amidst the pressure of our competitive institutions." Now, after my experience, I am more apt to agree with Emerson who declined the invitation to join Brook Farm. "Perhaps it is folly," he said, "this scheming to bring the good and like-minded together into families, into a colony. Better that they should disperse and so leaven the whole lump of society."

That Old Devil Sex. If drugs are potentially dangerous to the creation of alternate communities, that "old devil sex" is scarcely less a problem. As members of group marriages and hip communes attempt to rearrange society's norms concerning monogamous or dyadic sex relations, they find new problems. In a study of group marriages, Larry and Joan Constantine maintain that "multilateral marriage seems to bring out, highlight, possibly even aggravate, problem areas in individuals and prior two-person marriages." [9]

In an ideal sense, marriage can be conceived of as a continual quest for intimacy, both of a sexual and emotional nature. It is obvious that the intensely emotional state we call intimacy is difficult for an individual to achieve with one other person. When intimacy with a number of people is desired, the possibilities for conflict increase geometrically.[10] Sex, as a form of intimacy, can release potentially destructive emotions. Obviously, this is not the case in certain hip communes, where sex is dispensed with cavalier directness. In other instances, unplanned changes in sex norms produce occasional group crises, not all of which can be remedied.

Phil Tracy cites the case of an architect named Alex, who, with his wife, Sunshine, joined a communal household venture with two other couples. Alex and Sunshine both eventually quit their professional jobs (Sunshine had been a social worker) and experimented with psychedelic drugs. Their original communal arrangement simply involved the sharing of food bills, household utensils, and stereo equipment. Each couple shared a private bedroom, and while the group shared trips of the literal and drug variety, sexual intimacy was not discussed. However, several months later

9. From a plenary address presented before the Indiana Council on Family Relations.

10. It may be true that happiness, security, or pleasure increase greatly in multilateral or group marriages, although this is difficult to measure at present.

> . . . while Alex and Sunshine were lying in bed after making love, Sunshine confessed that she and Bob felt a deep attraction to each other and wanted to sleep together. She didn't want to hurt Alex, and if their marriage was going to be threatened by it, she and Bob would stop spending time together; but their relationship had reached a point where making love was only a natural progression in their involvement with each other. She wanted Alex to know. . . . Alex and Sunshine discussed it well into the dawn. Alex felt his whole world caving in around him in the bed and could not bring himself to make love to Sunshine before they collapsed into a fitful sleep, even though she begged him to.[11]

Both Alex and Sunshine finally agreed to submit their problem to a discussion of the entire house, and after some days of emotional discussion, the group decided that couple relationships were "mutually restrictive and repressive." A short time later, Bob and Sunshine made love.

> Nothing happened right away. Sunshine and Alex continued to sleep in the same room, but a gentle chasm slowly separated them. Alex became fascinated with sexual techniques which up to then he had considered bizarre. Sunshine, sensing Alex was competing with Bob for her ultimate intimacy, resisted. Their love making became less frequent.[12]

Later, other members of the group began to lose their sexual exclusivity as well. Finally, the group developed the idea that couples should no longer share rooms. Phil Tracy reports that at this point,

> A strange distance developed among the family as a result of the change in living arrangements. While previously much of their free time was spent in group activity, now they came together only during meals and house meetings, pairing off at other times in arbitrary but limited couples. . . .[13]

Many groups, such as the Mormons, have shown that experiments with sexual and family life can work to a degree, at least. However, with insufficient rules to guide one's behavior and no concept about the basic nature of man and woman, many experimenters find their

11. Phil Tracy, "Life In A Commune: A Fable," *National Catholic Reporter*, December 4, 1970, p. 13.
12. *Ibid.*
13. *Ibid.*, p. 14.

trial and error methods of family reorganization demoralizing to the group. Just as the use of drugs can wipe out group consciousness by limiting individual's perceptions to their own private world, sex relations can fragment group consciousness into competition-oriented "pair consciousness." When several individuals compete for sexual favors, individual egotism supersedes concern for the group. If, then, the group's survival is threatened, only two logical alternatives remain. Either rules must be set up to restrict sexual competition (marriage would do this), or sex must be somehow separated from intimacy so that it becomes another purely physical need. The latter option is, of course, extremely difficult for people in this society to accept. So thoroughly have we been socialized with the idea that sexual contacts entail responsibility and possession, that severe guilt often results from casual sex.

If intentional communities and hip communes survive their crises of leadership and the interpersonal problems of drug use and sex, they still must question their *raison d'etre*. Most of the modern communalists I talked with were extremely skeptical of dogmas and arbitrary rules. This is probably psychologically healthy. Yet, while dogmatic theology or ideology produces intellectual prisons, complete avoidance of ideology produces chaos. What seems to be needed is an orientation that gives people a focus of dialogue, or a common concern, without obviating individual differences. Many of the religious communes described in this book have achieved this. Several of the Walden-Two communities seem to be working toward this end as well. The hip communes that survive will probably be those that have developed a focus of concern other than their sex and drugs. Common concerns could revolve around the arts or crafts, voluntary primitivism, or hip service work such as that practiced by the Diggers. Needless to say, communes must encourage intellectual and creative growth in order to survive.

11

THE PROMISE OF COMMUNAL LIVING

"That communist societies will rapidly increase in this or any other country, I do not believe. The chances are always great against the success of any newly formed society of this kind. But that men and women can, if they will, live pleasantly and prosperously in a communal society is, I think, proved beyond a doubt. . . ."

Charles Nordhoff, 1875

"Some pass beyond good and evil and see only the play of primal and exuberant energy; some fall in love with the surface of things and make beauty into a theodicy; some play the game of life, in and out of it at the same time. This is the play attitude toward life."

George Kateb

CRITERIA FOR SUCCESS

Discussions of communes usually include theories about the causes of community failure or success. What benchmark can be used to measure communal success? Generally, critics measure communal success in three ways. The first concerns the effect communal living has on the larger social order. If communes can significantly change society for the better, they are worthwhile—that is to say, successful. It is, of course, difficult to perceive the subtle ways that communes have changed America. To be sure, early American communalists were among the first to call for racial and sexual equality. In 1835, the Owenite communalists donated their New Harmony Library to the Academy of Natural Sciences at Philadelphia. This is sometimes cited as the first truly public library in America.

Other writers have shown the impact of communal and utopian thought on education and city-planning. W. H. G. Armytage argues that "community experiments . . . are among the most important and universal ways in which societies the world over have maintained their vitality and advanced in type." [1]

If we were to conceive of the communal life-style as a teacher, however, and the rest of the nation as a pupil, it would soon be evident that the pupil has little motivation to learn. Admittedly, communal living can point to the advantages of nonexploitive, cooperative living. However, moralistic or ethical appeals have a limited audience. Evidently, many powerful individuals in our society have vested interests in continued international tension, economic exploitation of slum-dwellers, and the mindless destruction of natural resources. These problems are, indeed, political and so gigantic that they are not likely to be solved by communalists, no matter how well-intentioned or noble. If, then, we were to judge communal organizations by their impact on the Leviathan that is American society, large-scale success could not be predicted.

Yet, it is perhaps unfair to judge communes by their immediate effect on the body politic. A more manageable definition of communal success pertains to the stability and endurance of the commune. Obviously, a majority of communes last only one or two years. Moreover, "successful" (long-lasting) communes have been known to cripple individuality, creativity, intellectuality, and sexuality. Can one for example, describe the rigorous and stultifying lives of the nineteenth-century Shakers or Rappites as successful? If stability in human relationships is success, prisons are among the most successful of institutions.

The most meaningful definition of communal success, it seems to me, relates to individuals rather than the larger society or group longevity. How are individuals' lives affected by their communal living? It is here that judgments must be made. It is also here that communalists stand the greatest possibility for success.

THE INTEGRATION OF FANTASY

Communalists are full of dreams, and, unlike many middle-class men and women, they are attempting to live out those dreams. Most men and women escape the tedium of their lives through mass-produced

1. *Heavens Below: Utopian Experiments in England* (Toronto: University of Toronto Press), p. 16.

distractions. Men who have never thrown a ball spend large portions of their free time watching other men play professional football or baseball for profit. Men with dull sex lives read pornography. The situations in a pornographic novel, or in professional football, are so far removed from the reality of the average American male's life that they are mere opiates. How many housewives, bored and angered by their routine of endless drudgery, find themselves watching interminable soap operas? Besides the agonizingly poor acting endemic to "the soaps," they require no creativity or intellect on the part of their watchers. Indeed, it could be argued that if pornography is defined as a work having "no redeeming social value," many of "the soaps" could be labeled obscene.

This gap between the reality of everyday responsibilities and one's dreams Kenneth Keniston calls the "disassociation of fantasy."

> When imagination and life are separated, imagination continues to operate but becomes sterile and escapist, no longer deepening life but impoverishing it at the expense of another dream world that contains all that real life lacks. . . . Our shared fantasies are almost entirely contrasts or oppositions to daily life: they contrast with the lack of violence or intense passion in the average man's life, and with the specialization and abstraction of his work. Thus they seldom serve to enrich life, but rather to vitiate its imaginative vitalities.[2]

The tragedy here is not that most men can never fulfill their dreams but that their dreams lose meaning as they mature into adulthood. Contrast Keniston's description of adult fantasies with that of the child.

> The child's imagination operates on the objects of his daily environment, which become mythologized—first invested with supernatural attributes and later loaded with a cargo of associations and mental images that add to the meaning of the "realistically perceived" object. . . . The games of children—played with utter seriousness—illustrate the same closeness of fantasy and reality: the child can work through the problems of growing up, the difficulties of associating with other people, and can learn to master his environment in a setting not at all detached from the other activities

2. Kenneth Keniston, "Alienation and the Decline of Utopia," in *Varieties of Modern Social Theory*, edited by Hendrik Ruitenbeek (New York: E. P. Dutton and Co.), p. 86.

of his life. Work, play, love and fantasy are closely intertwined. . . .

When communalists blunder or falter in their activities, they still retain pride in having tried to live out their lives in consonance with their visions.

Patsy (Richardson) Sun describes her communal fantasies and her attempts to live them in the following section. Patsy, together with a dozen other hardy, hip communalists, lived out her dreams in a small farm in northern Minnesota. Freefolk, Patsy's communal experiment, lasted somewhat over a year. In some ways, of course, Freefolk did not live up to its expectations, but in another sense Patsy Sun sees her life as more human and richer because of her days at the farm.

WE LEARN BY DOING

by Patsy Sun

The night we first got here, hitching in from the closest bus depot in late November last year . . . it was snowing and there were no lights on, and the driver of the car who had gone out of his way to drive us out to the farm asked for the tenth time, "You sure this is the place?" He couldn't believe we had come all the way from Connecticut with almost all our worldly goods on our backs to move to a place we'd never been to, live with people we didn't know, by god, weren't even related to! We fumbled in the dark to light the kerosene lamp, and there were all the remembered smells of wood smoke and linoleum and maybe sauerkraut fermenting. I think Bob built up the fire, and I got out my guitar, and we waited for a while, looking at the pictures on the wall, checking the place out the way you do when the host is out of the room. Then they came in, Ferdi and Rebecca and little Geordie, these pacifist-anarchist types we'd come to live with on a hunch or a hope.

We stood around looking at each other, feeling shy. I remember Rebecca asked in one sentence whether we were organic gardeners, nudists, and anarchists. . . . It was a funny way to begin a community, if that's when it happened, standing around the stove looking at each other, without any outlined programs or objectives, not even knowing each other or what we meant by community. And I

3. *Ibid.*, p. 88.

think that it is a good thing we didn't write it down, but mostly let what happened happen.

For over a year now we've called ourselves freefolk, being freer than some and wanting to be more free. And people come and go and stay longer than they planned or come back, or maybe leave to search further. We'd like more people to share this life with us, who think their dreams might sit well with ours, people who are seriously interested in community as a way of life. There are no other membership requirements. We like visitors, too, or people who want to stay a short while to learn more about homesteading and community living, but we like them best when they write to us first to let us know they will be passing by and see if it is a good time to drop in. Living out here in the north woods your mind gets blown easily if you see too many strangers all moving in on you at once. When our numbers doubled in a few days this summer we panicked, thinking of other communities that had been swollen and burst by a sudden overwhelming population explosion. There were ten or twelve of us here then, I think, and we decided that what we wanted was a small close community of twelve or fifteen adults. It was a hard decision to make for people who wanted to share their possessions, their land, and their lives, to start writing letters, "Listen man, it is really great to hear you are interested in community, but we are swamped with visitors . . . and could you come in the fall." It was painful, and I'd write these complex replies trying to explain, and that is what we decided. We weren't going to throw anybody out, we'd just try to explain, kindly but firmly, how we felt . . . and so far it's worked. Someday, perhaps, after we have really established a community here, we will split and divide—mushroom and spread—right now we're growing too hard to think of it.

It was about that time we started having meetings when new people arrived to talk about their hopes and plans, to share our feelings and to explain what we were and what we wanted to be. It is sort of an artificial custom, and it has never come quite naturally, but for us it was a necessary way to avoid misunderstandings and let us all know what was happening.

We talked about drugs, too, and where we all were with them. Too many places had been busted. At other communities we knew of, heads got paranoid, avoided local residents, only wanted hip people around. For us, part of the whole scene was a kind of witness to the life we believed in. We didn't want to start excluding straight visitors or playing cops and robbers (who's the narc?). There were a lot of other drug communities around: we wanted to try something different. So the word spread, and we helped it, "They don't have any rules there, but don't bring drugs."

One of the first things people ask us is "So how do you get along with your neighbors; do you get a lot of static?" and they are always surprised when we say, "Great, it is one of the best things about living here." A community surrounding our community has developed, partly because we are living in an area where money is scarce. The divergence in ways of life isn't that great. Bartering, trading, exchanging labor is a frequent occasion. We are able to help some of our neighbors, and they often help us, giving us the skim milk they would otherwise throw out, teaching us stuff you won't find in books. Another major reason we have a good thing going with our neighbors, besides the fact that we really dig people and try to make friends, is that the community has grown slowly, giving people a chance to get used to us, to know that we are friendly, "hard-working, honest folk, even if they do look a bit weird." Rebecca and Ferdi were here alone for two years before anyone came to join them, which was too hard on them to be recommended as a really good idea, but there is no doubt that it has helped our local relationships. This doesn't mean that there aren't rumors—but there are enough people who know and like to offset them.

There are eight of us here now, counting the kids (and kids do count, making more noise and taking up more room than the rest of us). Deena came here with her twin babies Josh and Amy, one and a half, looking for a place to grow roots. She lives upstairs in the loft room. Bob and I have built a cabin down in the woods under the fir trees. Ferdi, Rebecca and Geordie sleep in the cabin across from the pump house. The community room, a lean-to on the barn, is where we all eat, sing, read, and bounce the babies.

All of us are city people; we learn by doing. With a book in one hand, seeds in the other, and a kind of optimism, we grow most of our own food. We have no electricity, no telephone, no running water, no tractor. We do have a car and a truck that runs sometimes, but not so many bills and as little dependency on the industrial-technocracy as we can now manage. We dig living this way—but it's more than that. For me, it seems like the whole industrial thing is based on exploitation. Exploitation of earth and trees and water, or animals and people and nations. It is something I don't want to be a part of, not raping the land or taking slaves either, even willing ones (to stand on an assembly line screwing on bolts is a job I couldn't bear for a week). Maybe technology could be used well, but it isn't now—and it is more than that.

For it seems machines separate me from a part of life I don't want to be separated from. I like to pump my water, to know that the power comes out of my own body. I like to drag wood in from the forest and to cut it by hand with a two-man saw, because it keeps me in touch with what I think is basic: food, warmth, water. So we live here, and we don't go to war or pay federal taxes. We encourage other people not to either. When the spirit moves us, we crank out leaflets, speak at meetings, or leave cryptic notes on the bulletin boards at the nearby state college. What all that means is we groove on people getting free. For every day is a kind of demonstration of what we believe.

Freefolk community may not be intentional, but it is on purpose. Each of us comes here, no matter how long he stays, for his own reasons; a lot of them are the same, but some things are more important for some people. For us at this point, this is a satisfying way of life, but it is more than an end in itself; it is a means toward a gentle revolution where people drop out of the system and take responsibility over their own lives. We want to create a place where work is fun and meaningful, where children can grow straight and beautiful according to their natural inclinations. A place where people have a chance to discover who they are and what they really want, where each man deals with his neighbor sensitively and compassionately, where laws and prisons and armies are not needed. We don't have any group goals. We dream of having a school here, or a camp, or trying some street theatre, of bringing Kountry Kulture to our local township meeting house.

It is hard to talk about community. There are so many things that don't make sense unless you have tried it. It is hard to talk about Freefolk because it is something different to everyone, so much of it, smells and sights and sounds. Whatever you say comes out sounding like not quite the truth. We haven't created a utopia here, I can say that. There are a lot of hassles. People who want to escape hassles should forget community. You get hurt a lot. And there is always the question whether the hurt is worth the good times, or whether you're learning anything from it. For me,

I guess so far it has been worth it. Every hassle is so basic it becomes a discovery in human need, your own and your friends.' Often I don't feel myself growing on beyond the insight, but maybe, I think, the insight itself is a kind of growth.

Harder than the hassles with each other are the hassles with yourself, wondering if this is, after all, where you should be, discovering that the world you wanted to create doesn't come all that easy, not even when people want it to. There is bitterness and maybe yelling. Sometimes there is silence which can be the worst of all, or hate; but more loving, I think, because we have the trees and the sunset fields to run in. Whatever mess our lives are in, the mess is at least our own mess and when we sing "Love means each other everyday," we know where that's at.

CREATIVITY, PLAYFULNESS

Lord Bertrand Russell was fond of the idea that two basic impulses guide man's ethical and political actions. The first, possessiveness, results in the desire to obtain and protect property. More often than not, possessive individuals use the apparatus of the state to their own material advantage, by exhorting the patriotic use of force to extend or protect their properties. The use of force is legal when used by the rich against the poor; extralegal when the poor "take the law into their own hands." Although Russell saw production and industry as potentially good forces for mankind, it is easy to see that the possessive impulse is potentially destructive to mankind.

"The creative impulses," Russell argued, "unlike those that are possessive, are directed to ends in which one man's gain is not another man's loss. The man who makes a scientific discovery or writes a poem is enriching others at the same time as himself." [4]

It follows from Russell's analysis that many restraints should be placed on the often hate-producing, possessive impulses. Few, if any, restrictions need be made on man's creative drives, since they do not threaten the social fabric or cause extremes in poverty and wealth.

Most of the thousands of American communalists would agree with Russell. A great majority would also agree with Russell's contention that there is no great virtue in work.[5]

4. *Political Ideals* (New York: Simon and Schuster, 1964).
5. See Russell's *In Praise of Idleness* (London: Unwin Books, 1935). Russell proposes a four hour work day with leisure time spent in creative loafing, such as painting or fulfilling one's scientific curiosity.

As Dick Fairfield, editor of the *Modern Utopian* puts it,

> We live in a capitalistic, overly competitive, profit-oriented, greedy society based on affluence and mass consumption—spending more and more dollars on more and more goods which we are convinced (by Madison Avenue, et al.) that we need. And the more you and I earn, the more we spend and pay in taxes to support this system, insensitive to the real needs of the rest of the world. The key to living a moral, pleasurable life is to reduce one's intake and outgo of material goods—to deal with the poverty of the world by rejecting America's materialistic standards—to deal with the negativity in the world by being as high, as joyous as possible in your daily life.[6]

Communalists, in so far as they are able to reject excessive materialism, come to see work as a game. For that reason, they thrive on work which might be considered degrading to a more sophisticated individual. Hip Hog Farmers made a beautiful game out of the care and feeding of piglets. It was play.[7] Other communalists raise chickens. They can tell you about Danish Brown Leghorns, Mottled Anconas, Silver Laced Columbians, Rhode Island Reds, or even Green-sheen Black Langshands. The fact is that in a society of affluence, playing at work can produce enough for survival. The idea of work as a game means that, in a sense, hip communalists have regressed to childhood. Children play at work. In this way, the young communalists have given a new meaning to the Biblical dictum that, "whosoever shall not receive the kingdom of God as a little child shall in no wise enter therein."[8]

Communalists are in a very real way seeking the kingdom of heaven on earth. Heaven is represented as a playful, joyous, and total way of life. The creative life sought by the communal experimenters may be beyond their reach. Nevertheless, no religious pilgrims or financial entrepreneurs have had more fun in their quests for unattainable goals. Then, too, hip communalists today can point to real and concrete achievements in their playful work. Architectural such as domes, zomes, and other exotic homesteads are monuments to the creativity and, yes,

6. Dick Fairfield, "We Are All One Vision," *The Modern Utopian* 1, no. 2 (January, 1970).
7. J. Huizinga's *Homo Ludens* (Boston: Beacon Press, 1955) is an excellent source on the philosophical import of play.
8. Luke 19:17.

skills of hip communalists. Planting gardens for the first time and the wonder of the greening process are creative miracles for the back-to-nature communalists.

Just as hip communalists play with nature in an essentially non-destructive way, the new Walden-Two communalists wish to play with technology. According to the Walden-Two philosophy, machines can be the slaves to give men and women the leisure they need to contemplate and create.

One of the important differences between hip or Walden-Two communalists and the *haute-culture* leisure classes is that the communalists do not derive their leisure from the daily toil of the masses. Unlike the international "jet set," whose conspicuous consumption is legendary, the new communalists gain leisure by simply cutting down their needs. It is true that some semiserious communalists get food stamps or other relatively small forms of the dole, but these amounts are infinitesimally small compared to the subsidies given certain giant American corporations.

It is sadly true that most of the world's population can not afford to be playful. Where poverty and war exist, childhood playfulness is lost. For this reason many of the world's poor see nothing amusing about the antics of American communalists. A playful society can exist only in the midst of plenty, and it seems likely that only serious revolution will make the lives of the masses tolerable in Latin America, Asia, or South Africa.

The playful creativity of young communalists is not at all like the creative play of intellectuals employed by the Rand Corporation or the Defense Department. The creative intellectual play of individuals such as Herman Kahn has little to do with the creation of human communities. In a direct and literal sense, the scenarios created by Kahn and others deal with the *destruction* of communities and nations. In his *On Thermonuclear War,*[9] Kahn coolly surveys the possibilities of creating a doomsday machine to destroy all mankind. Kahn does not advocate the actual construction of such a diabolical weapon. He is merely playing with the idea. In another work[10] Kahn proposes that the term "wargasm" be given the ultimate nuclear conflict. Thus, it would seem that while young communalists are playing at community building, representatives of the military establishment play with ideas estimating

9. Princeton, N.J.: Princeton University Press, 1960.
10. *On Escalation: Metaphors and Scenarios* (Baltimore: Penguin Books, 1968).

the blast-damage, thermal radiation, and fallout of yet-unfought, ultimate wars.

History will judge the relative humanness of the long-haired communal dreamers and the well-dressed visionaries of hypothetical and real wars.

ACCEPTANCE OF ONE'S BODY

One of the more interesting theories concerning the disgust or even hatred of "hippie freaks" by "middle Americans" has to do with human hair. Why are long-haired, white students and bushy "Afro'd" blacks seen as such a threat to the social order? Why, in 1970, were school principals measuring the hair length of defiant and shaggy students? One facile answer is that outrage is vented on the longhairs chiefly by those whose follicles have given up. While this is a somewhat zany theory, it does point to a real difference between generational groups. Middle-aged Americans spend many hours and much money to control, straighten, curl, dye, deplete, delete, and deodorize hair. This is to say nothing of two new "hair terrors" discovered in the late 1960s—"split ends" and "the frizzies." It may well be that a Martian who knew of the American people only by watching television would conclude that the greatest problems faced by the humans were those of ill-smelling breath and "itchy, flaking dandruff."

Beyond the needs of sanitation, Americans spend hundreds of millions of dollars to control areas of their bodies which may give offense to others. Corporations exploit these anxieties to good advantage and produce a really neurotic concern with the potential ugliness, smelliness, or offensiveness of the human body.

The answer hip communalists give to this phenomenon is quite clear: the human body is good and natural, complete with hair and the rest. Most hip communalists (at least in my experience) do bathe. It is true that examples of uncleanliness and disease can be cited for several hip communes. Yet, for the most part, hip communalists do take reasonable precautions against disease. Beyond that, however, few care which direction their hair is growing at a given time.

The body, according to the hip point of view, is a celebration. It is not to be hidden, twisted out of shape, or confined. Hip communalists, like their ideals, the Indians, love to dance. The dance is a celebration of the body as well as an appreciation of movement and rhythm. The ecstasy of the dance, together with nudity and/or wildly expressive clothing, seems to provide an awareness for communalists that the body and mind are not separate. In short, hip communalists are not alienated from their bodies. Although, some communalists abuse their bodies (with drugs and improper dieting, for example), it is senseless to condemn the majority of communalists, who have discovered the truth of the Biblical statement "Man is that he might have joy."

Structured religious communes, especially those of an Eastern mystical variety, spend a great deal of their time with exercises such as *hatha* yoga or *za-zen*. They, too, have come to understand that the body is more than a machine to perform work. Cultivation of body control is spiritually enlightening as well as good in its own right, according to their view.

Since a vast majority of the new communalists are young, it is impossible to judge whether older people in a communal situation could learn to "groove on their own bodies" in the same way that the young do. In fact, it may be that communalists accept their bodies because of their youth rather than their living arrangements. Fewer and fewer American young people are taught to perceive their bodies as a source of sin and shame. Perhaps, in this way, the communal movement is partly an extension of trends in the larger society.

TOWARD COMMUNITIES OF TRUST

The word "sharing" may have unpleasant connotations to anyone who remembers childhood quarrels with brothers and sisters. Parents, nearly always scrupulously honest in dividing candy or cake, functioned as referees in contests to see who got the biggest piece. The young communalists described in this book grew up with precisely the same childhood frustrations, rivalries, and jealousies as did other Americans. Yet, somewhere along the line, they became aware that fulfillment does not lie in the incessant accumulation of things. "People," one communalist told me, "are the ultimate trip." When material goods are not given crucial value, people can relate to people with no fear of exploitation.

Nearly every hip commune has its parasitical hangers-on. Unless the community is fighting for its existence, no one seems to care much. Food, clothing, dope, cars—they really are not important except that they can be used for play. Possessiveness, like lust, is a deadly serious business. Hip communalists have little time to be serious, except, of course, when their existence is threatened. Some accuse the new communalists of theatrical intentions in their sharing and lack of concern for their material possessions. This may or may not be true, since in reality no one can "look inside the heads" of the young experimenters to perceive their "true" motivations. In point of fact, it may not matter if the sharing displayed by communalists is "an act" with ulterior motives. If they continue to share with each other, their motives often change.[11]

What is important is that the new communalists are people-centered rather than thing-centered. Unlike the people-centered philosophies of Dale Carnegie and Norman Vincent Peale, "success" or instant popularity is not the goal of most of the young drop-outs. They are not interested in "learning to get along better with people" in the Carnegie method (rule number 13, SMILE!). Honesty and depth of feeling are, in the end, far more important than manipulation for personal gain. Hip communalists are honest, sometimes humorously so, sometimes ruthlessly so ("Max, get your butt down to the creek and wash. You smell like a goddamn sewer.")

Because of their concern for equality and their antibureaucratic views, hip communalists need not fear offending the powerful or someone who will withhold official favors from them. It is basically this lack of fear that gives the new communalists the ability to be more honest and open in their dealings with each other.

More than this, today's social experimenters have learned something about the nature of authority. It is this: no man should respect another as an authority simply by virtue of tradition or bureaucratic stature. The creators of the alternative life-styles have little respect for those whose authority is dependent on favorable kill-ratios, gigantic profits, or manipulation of voters by good public relations. Authority is, of course, needed for any group to survive, but when it is based on technical competence at the expense of humane values, it should be rejected. For man to adapt to his society and to his environs, a system of authority or polity must be created. The young counterculturalists are beginning to realize this. Those who decry the young for losing respect for authority, fail to see that much authority is arbitrary, repressive, or based on force. New forms of authority are hopefully developing in the communal scene—authority revolving around self-insight, knowledge of group dynamics, and cooperation with nature. As to those in the larger society

11. This idea is explored by G. W. Allport in "The Functional Autonomy of Motives," *American Journal of Psychology* 50 (1937): 141–156.

who seek control over other men and materials, Harold Lasswell's analysis is probably correct: "Power is expected to overcome low estimates of the self." [12] In other words, powerseekers are constantly attempting to compensate for personal deprivation. Although most communalists have little formal training in psychology, they seem to intuitively realize that fact.

In many ways, the new communalists, whether they be of a religious, hip, or Walden-Two variety, seem to have regressed to the status of children. After all, they seek a unified, simple, often magical world view, much like that of children. Their sex roles are blurred. They love to play. They are sometimes irresponsible. They have not yet learned to be cynical about their roles in their new groups.

Is it of necessity wrong to wish to hold on to childhood and to resist adulthood in American society? "And now," mused Harry Stack Sullivan, "when you have ceased to care for adventure, when you have forgotten romance, when the only things worthwhile to you are prestige and income, then you have grown up, then you have become an adult." [13]

The next decade in America is likely to see a great proliferation of communal ventures founded by "the gentle people." Those of us with humanistic values should applaud the few groups that manage to survive the rigors of communal pioneering. For the large majority of groups that can be expected to fail, one is reminded of the advice given Charles Nordhoff by a member of the Icarian commune. It was 1878, and the French Icarians were on the brink of defeat after nearly forty years of utopian socialism in the new world.

> Deal gently and cautiously with Icaria. The man who sees only the chaotic village and the wooden shoes, and only chronicles, those will commit a serious error. In that village are buried fortunes, noble hopes, and the aspirations of good and great men . . . .[14]

The only way to fully appreciate the communal movement is to participate in it—not blindly or with complete naiveté, but with the

12. *Power and Personality* (New York: Norton, 1948).

13. Quoted in Thurmond Arnold, *The Folklore of Capitalism* (New Haven: Yale University Press, 1937), p. 163.

14. Charles Nordhoff, *The Communistic Societies of the United States* (New York: Harper and Brothers, 1875), p. 339.

knowledge that human institutions are temporal and finite. How does one begin? Patsy Sun has a plan.

A GRAND MASTER PLAN

by Patsy Sun

to build a world without fear or hatred
to share one's life and livelihood
to become what one really is
to find the human way

Step one: Go somewhere where no one else that you know is. Buy some cheap land, a copy of *Organic Gardening Encyclopedia*, and some seed. Establish a base camp disguised as a self-sufficient farming community.

Step two: Make friends with the local farmers. Ah, good people! They don't have much, but they'd share that. Always they give us more than we can return, but like one neighbor says, "What's a few pumpkins between friends?"

Step three: Infiltrate the local peace group. And good people they . . . come to visit us with electric coffee percolator. "Where's the plug?" We heat it on the wood stove and talk 'til midnight. Later they invite us into town to speak on panel: "Americanism in the '60s" or "Civil Disobedience." Turn people on to doing things for themselves. ("Stop bitching for better schools or housing or welfare. Go out and build them. Stop paying taxes for war. Refuse the draft.") Turn people on to community and living simply. Turn them on and see the light go click behind their pale eyes. ("But you just can't live on an untaxable income!" "But man, we're doing it!" . . . much laughter.)

Step four: Make friends with heads and friendly students at your local teachers' college. (Even up here there is a psychedelic shop. Under a rock and behind the trees come a few draft resisters, a poet and a folksinger, too.) Get them to set up a draft table at the college. Get almost thrown in the lake by the campus veterans. Retreat sometimes . . . but return again. Invite people out to the farm. Turn them on to milking a cow. Turn them on to being turned on without drugs. Let them turn you on with their music. Look at each other and smile a lot . . . who can help it!

Step five: Make friends with your local feds. (Ah, not so good people!) Entertain them when they arrive to ask you how come you're writing all those letters to the draft-eligible men in the country or how come you're not in the army or how come you aren't married to the girl you live with. Offer to show them the cow. Or offer to show them the door. But remember their names. They will most likely come back. . . .

Step six: Drop ideas on peoples' heads instead of bombs. Swoop into Duluth for a conference at the U. Let an audience capture you and spread the word. We're Free! You're free! All you have to do is do it. Whatever bonds hold you are tied with your own hands. And you don't have to go to college for four years, or get a "good" job, or get married, or cut your hair short, or wear a girdle or join the army, or pay taxes. You may pay a price, but then there is a price for everything. Whatever it is that you really want to do, do it now, for life is short and love is fleeting when it's not spent. Meet new draft resisters. Love them all. Sing. Talk. Drink wine. Invite them out to the farm for a week when school's out. Then retreat.

Come home. Dig your toes in the warm dirt. Pick a tick off your friend's neck. Have a few stupid arguments. Write to the urban poor telling them you'd like to help families get out of the city if they want. Go out and plant a row of carrots. Make a mistake. Roll in the grass and begin again.

INDEX

K

L

M

N

O

P

Q

R

S

T

V

W